MW01322348

Careers in Focus

ARMED FORCES

Ferguson
An imprint of Infobase Publishing

Careers in Focus: Armed Forces

Ferguson
An imprint of Infobase Publishing
132 West 31st Street
New York NY 10001

Library of Congress Cataloging-in-Publication Data

Careers in focus : Armed Forces.
p. cm.
Includes bibliographical references and index.
ISBN-13: 978-0-8160-7288-0 (alk. paper)
ISBN-10: 0-8160-7288-4 (alk. paper)
1. United States Armed Forces—Vocational guidance—Juvenile literature.
I. Ferguson Publishing.
UB147.C37 2008
355.0023'73—dc22

2008018311

Ferguson books are available at special discounts when purchased in bulk quantities for businesses, associations, institutions, or sales promotions. Please call our Special Sales Department in New York at (212) 967-8800 or (800) 322-8755.

You can find Ferguson on the World Wide Web at http://www.fergpubco.com

Text design by David Strelecky
Cover design by Salvatore Luongo

Printed in the United States of America

Sheridan MSRF 10 9 8 7 6 5 4 3 2 1

This book is printed on acid-free paper.

Table of Contents

Introduction 1
Accounting, Budget, and Finance Occupations 7
Art, Design, and Music Occupations 14
Aviation Occupations, Except Pilots. 25
Combat Specialty Occupations. 34
Construction, Building, and Extraction Occupations. 45
Counseling, Social Work, and Human Services Occupations 54
Engineering, Science, and Technical Occupations 63
Health Care Occupations. 76
Information Technology and Computer Science Occupations 90
Intelligence Professionals 97
Interpreters and Translators 105
Law Enforcement, Security, and Protective Services Occupations. 113
Legal Professionals and Support Occupations. . 122
Mechanic and Repair Technologists and Technicians 134
Media and Public Affairs Occupations. 146
Military Pilots 155
Military Recruiters. 164
Naval and Maritime Operations Occupations . . 172
Personal and Culinary Services Occupations . . 179
Transportation, Supply, and Logistics Occupations 187
Appendix: Monthly Basic Pay by Pay Grade and Experience. 197
Index 199

Introduction

Five separate military services make up the U.S. armed forces: the army, navy, air force, Marine Corps, and Coast Guard. These branches organize, train, and equip the nation's land, sea, and air services to support the government's national and international policies. Together, they are responsible for the safety and protection of the citizens of the United States. Those who choose to be members of the armed forces dedicate their lives to protecting their fellow Americans.

The army is the senior service. It is traditionally known as the branch that fights on land. Most of the United States' 25 million living veterans served in the army. The army continues to be the largest of the services in total recruits, with 433,000 enlisted soldiers and 85,000 officers and warrant officers.

The navy, more than any of the other services, has a special way of life. Guided by traditions of the sea, this branch is in many ways more of a closed society than the other services. Its officers and enlisted people work and live together at sea for long periods—a lifestyle that demands close attention to duties and teamwork. Ships and aircraft units visit many parts of the world. It can be an unusual and wonderful life, and strongly appealing to many who are looking for a different and exciting type of career. The navy has about 281,000 enlisted personnel and 51,000 officers and warrant officers.

The air force, newest of all the services, is highly technical and appeals to those interested in aviation and mechanical trades. Approximately 263,000 enlisted personnel and 65,000 officers are employed by the air force.

The Marine Corps operates on land and sea, and marines usually form the advance troops in military operations. The corps is closely associated with the navy, and like the navy, prides itself on meeting the highest possible standards in training, military bearing, and discipline. Apart from more military duties, marines provide security on navy property, and guard U.S. embassies and consulates around the world. Approximately 167,000 enlisted soldiers and 20,000 officers and warrant officers serve in the marines.

The Coast Guard is the smallest of the military services, and, as such, offers unique opportunities. It is responsible largely for the enforcement of maritime law, but is perhaps most well known for its involvement in search and rescue efforts, aiding those in distress at sea. Although opportunities exist for overseas assignments, most duties in the Coast Guard are related to the waters and shores of the United States. Currently there are 42,000 enlisted members, warrant officers, and commissioned officers in the Coast Guard.

Positions in the military fall under two broad occupational categories: *enlisted personnel* and *officers*. Enlisted personnel execute the daily operations of the military, and are considered noncommissioned officers. Commissioned officers function as managers of the military, overseeing the work of the enlisted personnel. (Note: Each career article contains a sidebar that will tell you if a career is for enlisted personnel or officers, and which branches offer this career.)

In addition, the army, navy, Marine Corps, and Coast Guard maintain a third classification of skilled experts called *warrant officers*. Enlisted soldiers or civilians who demonstrate technical and tactical ability in any one of several dozen occupational specialties may qualify as a warrant officer. Warrant officers are highly specialized and gain additional expertise by operating, maintaining, and managing the specialty's equipment, support activities, and technical systems throughout their careers. Specialties include, but are not limited to, missile systems, military intelligence, telecommunications, legal administration, and personnel.

A broad general difference between the requirements for enlisted personnel and officers is academic preparation. The service branches accept applicants of varying ages and educational backgrounds, although officers are required to have college degrees and enlisted people are not. Those who intend to serve as enlisted personnel should finish high school and then enlist. (To enter the military, you must be at least 17 years old. Applicants who are age 17 must also have the consent of a parent or legal guardian.) High school graduates are more likely to be successful in the military than nongraduates, and the services accept few applicants without a high school diploma. All service applicants must take the Armed Services Vocational Aptitude Battery exam as a requirement for enlistment. You must also sign an enlistment contract. This is a legal agreement that will bind you to a certain amount of military service, usually eight years. Active duty comprises two to six years of this agreement, and the remainder is normally spent in the reserves.

All service personnel receive special training in military skills. Those who want to rise in the ranks have the opportunity to attend school or undertake independent study and will be rewarded with advancement. Military education in many instances is related to civilian occupations, an added incentive for those who later decide not to make a career of the military.

While many recruits enter the armed forces immediately after graduating from high school, the military uses every opportunity to advance the education of its recruits. Frequently, service members in both categories, officers and enlisted personnel, obtain undergraduate and advanced degrees on their own initiative while in the service. The range of educational possibilities is diverse and often exciting, and financial help for study at outside institutions is readily available. You can learn to operate the engineering equipment of an aircraft carrier or, with appropriate academic qualifications, proceed through medical training to become a nurse. With a serious sense of commitment and a great deal of rigorous preparation, you even can join the navy's SEALS program, or "get jets," and become a member of the Blue Angels.

Each military branch has nine enlisted grades and 10 officer designations. The names of ranks vary among the services. This is why a simple numbering system has been adopted to denote rank. Promotion depends on ability, number of years served, and length of time since the last promotion. On average, a diligent enlisted person can expect to earn one of the middle noncommissioned or petty officer ratings; some officers can expect to reach lieutenant colonel or commander. Outstanding individuals advance beyond those levels.

Military life is extremely regimented. Service members are not always able to choose their assignments, places of work, or homes. Their responsibility is to serve the country and the military unit, so the needs of the country and the service come first. The modern services try hard to give members what they want; a good worker gets special consideration for the same reason a civilian employer rewards a top employee. But most service members at some time will, for example, work on the East Coast when they would rather be on the West Coast, go to sea when they would rather stay home, live in a house or barracks on the base when they would rather have an apartment in town, or serve a year in a post abroad when they would rather be with their families.

In contrast to these personal sacrifices, it's heartening to remember that the country always will need a military, both for defense and to protect its interests and citizens around the world. It's true that over time, often because of political or economic influences, the requirements necessary to meet these needs change. As America's position in the world grew stronger in the early 1990s, for example, the military engaged in a major downsizing, reducing personnel in nearly every branch and at all levels. The size and strength of each service reached a steady state in the late 1990s and the armed forces had to combat a public misperception that military jobs were not as available as they had been in the past. However, opportunities for a career in the military services continued to be plentiful.

It is probable that the war on terrorism, begun after the attacks on the United States on September 11, 2001, will be a long, complicated effort engaging all branches of the military and law enforcement. In addition to military action, personnel will be involved in any rescue and assistance needed on domestic soil to combat threats of additional physical, chemical, or biological attacks.

Each article in this book discusses in detail a particular military-related occupation. The articles have been written using the latest information from the U.S. Department of Defense, the five military branches, the U.S. Department of Labor, professional organizations, and other sources.

The following paragraphs detail the sections and features that appear in the book.

The **Quick Facts** section provides a brief summary of the career including recommended school subjects, personal skills, work environment, minimum educational requirements, salary ranges, and employment outlook. This section also provides acronyms and identification numbers for the following government classification indexes: the *Dictionary of Occupational Titles* (DOT), the *Guide for Occupational Exploration* (GOE), and the Occupational Information Network (O*NET)-Standard Occupational Classification System (SOC) index. The DOT, GOE, and O*NET-SOC indexes have been created by the U.S. government. Readers can use the identification numbers listed in the Quick Facts section to access further information about a career. Print editions of the DOT (*Dictionary of Occupational Titles*. Indianapolis, Ind.: JIST Works, 1991) and GOE (*Guide for Occupational Exploration*. Indianapolis, Ind.: JIST Works, 2001) are available at libraries. An electronic version of the O*NET-SOC (http://online.onetcenter.org) is available on the Internet. When no DOT, GOE, or O*NET-SOC numbers are present, this means that the U.S. Department of Labor has not created a numerical designation for this career. In this instance, you will see the acronym "N/A," or not available.

The **Overview** section presents a brief introductory description of the duties and responsibilities involved in this career. Oftentimes, a career may have a variety of job titles. When this is the case, alternative career titles are presented. Employment statistics are also provided, when available. The **History** section describes the history of the particular job as it relates to the overall development of its industry or field. **The Job** describes the primary and secondary duties of the job. **Requirements** discusses high school and postsecondary education and training requirements in both the civilian and military sectors, and personal requirements for success in the job. Exploring offers suggestions on how to gain experience in or knowledge of the particular job before making a firm educational and financial commitment. The focus is on what can be done while still in high school (or in the early years of college) to gain a better understanding of the job. The **Employers** section provides information on employment in the military. **Starting Out** discusses the best way to enter the armed forces. The **Advancement** section describes the advancement system in the military. Earnings provides an overview of the military pay system and describes the typical fringe benefits. The **Work Environment** section describes the typical surroundings and conditions of employment—whether indoors or outdoors, noisy or quiet, social or independent. Also discussed are the stresses, strains, and possible dangers of the job. The **Outlook** section summarizes the employment outlook in both the military and civilian sectors. For the most part, Outlook information is obtained from the U.S. Bureau of Labor Statistics and is supplemented by information gathered from professional associations. Job growth terms follow those used in the Department of Labor's *Occupational Outlook Handbook*. Growth described as "much faster than the average" means an increase of 27 percent or more. Growth described as "faster than the average" means an increase of 18 to 26 percent. Growth described as "about as fast as the average" means an increase of 9 to 17 percent. Growth described as "more slowly than the average" means an increase of 0 to 8 percent. "Decline" means a decrease by any amount. Each article ends with **For More Information**, which lists organizations in both the civilian and military sectors that provide information on training, education, internships, scholarships, and job placement.

Careers in Focus: Armed Forces also includes photographs, informative sidebars, interviews with professionals in the field, and an appendix that lists basic military pay by pay grade and experience.

Accounting, Budget, and Finance Occupations

OVERVIEW

The military allocates billions of dollars for supplies, equipment, and payroll. Financial experts, such as *accounting, budget, and finance specialists and managers,* are needed to budget, organize, process, and record these expenses. Opportunities for both enlisted personnel and officers are available in the U.S. Air Force, Army, Coast Guard, Marines, and Navy.

QUICK FACTS

School Subjects
Business
Economics
Mathematics

Personal Skills
Following instructions
Leadership/management

Work Environment
Primarily indoors
Primarily multiple locations

Minimum Education Level
Varies by career specialty

Salary Range
$15,617 to $30,618 to $66,154 (enlisted personnel)
$25,358 to $70,877 to $174,103 (officers)

Outlook
About as fast as the average

DOT
160, 216

GOE
09.03.01, 13.01.01, 13.02.03

O*NET-SOC
13-2011.00, 13-2011.01, 13-2011.02, 43-3031.00

HISTORY

Accounting records and bookkeeping methods have been used from early history to the present. Records discovered in Babylonia (modern-day Iraq) date back to 3600 B.C., and accounts were kept by the ancient Greeks and the Romans.

Modern accounting began with the technique of double-entry bookkeeping, which was developed in the 15th and 16th centuries by Luca Pacioli, an Italian mathematician. After the Industrial Revolution, business grew more complex.

As government and industrial institutions developed in the 19th and 20th centuries, accurate records and information were needed to assist in making decisions on economic and management policies. Accountants and other budget and finance professionals were hired to help businesses and the government keep their financial records.

In the military, accounting, budget, and finance professionals have been needed ever since America's first army, the Continental

Army, was formed in 1775 to fight the British during the Revolutionary War. Supplies and weaponry needed to be purchased, and soldiers had to be paid. This has not changed through the centuries of war and peacetime operations. Today, military finance professionals manage billions of dollars that must be allocated for military use in the United States, as well as throughout the world.

THE JOB

Accounting, budget, and finance specialists organize and record financial accounts for the military. They are responsible for payroll, including computing and recording payments and issuing checks, keeping track of direct deposit accounts, and disbursing advance pay to certain personnel. Specialists must keep detailed reports of financial transactions and records. They periodically review departmental financial records to check for inaccuracies or discrepancies.

All financial dealings conducted by the military are under the direction and management of *accounting, budget, and finance managers.* They work to create and implement policies on how military money is to be spent—for example, for reimbursement for travel expenses or foreign providers of supplies and equipment during oversees deployments. They also develop ways to track transactions, and manage the work of financial and accounting staff members. They coordinate with other military department heads to work on current and future departmental budgets, accounting, and other financial issues.

REQUIREMENTS

High School

If you are interested in a career in this field, you must be very proficient in arithmetic and basic algebra. Familiarity with computers and their applications is equally important. Course work in English and communications will also be beneficial.

Rank and Military Branch by Occupation

Job Title	Rank	Military Branches
Finance and Accounting Managers	Officer	Air Force, Army, Coast Guard, Marines, Navy
Finance and Accounting Specialists	Enlisted	Air Force, Army, Coast Guard, Marines, Navy

Source: U.S. Department of Defense

Postsecondary Training

Educational requirements in the civilian sector vary by career. For example, a bachelor's degree with a major in accounting, or a related field such as economics, is highly recommended for accountants. Some employers prefer to hire bookkeeping and accounting clerks who have completed a junior college curriculum or those who have attended a post-high school business training program. In many instances, employers offer on-the-job training for various types of entry-level positions.

In the military, finance and accounting specialists train via classroom instruction. Workers learn basic accounting principles and procedures and how to prepare financial budgets and reports, interpret financial data, and compute pay and deductions.

As officers, finance and accounting managers must enter the armed forces with at least a bachelor's degree in their field. Once they join the military, they receive specialized training via classroom instruction. Courses provide an overview of military accounting techniques, financial management methods, and other topics.

Visit the U.S. Department of Defense's Web site, http://www.todaysmilitary.com, for more information on military training in accounting, budget, and finance careers.

Other Requirements

Financial and accounting specialists and managers excel at mathematics, statistics, bookkeeping, and accounting. They must be detail-oriented, and able to work with accuracy. Managers must be able to direct and supervise the work of team members.

Visit the U.S. Department of Defense's Web site, http://www.todaysmilitary.com, for more on personal requirements for workers in these careers.

EXPLORING

If you think a career in this field might be for you, try working in a retail business, either part time or during the summer. Working at the cash register or even pricing products as a stockperson is good introductory experience. You should also consider working as a treasurer for a student organization that requires financial planning and money management. It may be possible to gain some experience by volunteering with local groups such as churches and small businesses. You should also stay abreast of news in the field by reading trade magazines and checking out industry Web sites.

To learn more about career opportunities in the military, visit the Web sites listed at the end of this article.

EMPLOYERS

The U.S. government employs the military. No specific employment statistics are available for military workers in accounting, budget, and finance. Overall, 1.4 million men and women are on active duty and another 1.2 million volunteers serve in the National Guard and Reserve forces.

STARTING OUT

Contact a military recruiter to learn more about career options. Visit the Web sites listed at the end of this article to locate a recruiting office in your area. To start out in any branch, you will need to pass physical and medical tests, the Armed Services Vocational Aptitude Battery exam, and basic training.

ADVANCEMENT

Each military branch has nine enlisted grades (E-1 through E-9) and 10 officer grades (O-1 through O-10). The higher the number is, the more advanced a person's rank is. The various branches of the military have somewhat different criteria for promoting individuals; in general, however, promotions depend on factors such as length of time served, demonstrated abilities, recommendations, and scores on written exams. Promotions become more and more competitive as people advance in rank. On average, a diligent enlisted person can expect to earn one of the middle noncommissioned or petty officer rankings (E-4 through E-6); some officers can expect to reach lieutenant colonel or commander (O-5). Outstanding individuals may be able to advance beyond these levels.

EARNINGS

The U.S. Congress sets the pay scales for the military after hearing recommendations from the president. The pay for equivalent grades is the same in all services (that is, anyone with a grade of E-4, for example, will have the same basic pay whether in the army, navy, marines, air force, or Coast Guard). In addition to basic pay, personnel who frequently and regularly participate in combat may earn hazardous duty pay. Other special allowances include special duty pay and foreign duty pay. Earnings start relatively low but increase on a fairly regular basis as individuals advance in rank. See the appendix at the end of this book for detailed information on pay scales for the U.S. military. When reviewing earnings, it is important to keep in mind that members of the military receive free housing, food, and health care—items that civilians typically pay for themselves.

Additional benefits for military personnel include uniform allowances, 30 days' paid vacation time per year, and the opportunity to retire after 20 years of service. Generally, those retiring will receive 40 percent of the average of the highest three years of their base pay. This amount rises incrementally, reaching 75 percent of the average of the highest three years of base pay after 30 years of service. All retirement provisions are subject to change, however, and you should verify them as well as current salary information before you enlist. Those who retire after 20 years of service are usually in their 40s and thus have plenty of time, as well as an accumulation of skills, with which to start a second career.

WORK ENVIRONMENT

Accounting, budget, and finance specialists and managers work in comfortable and well-lit offices located on military bases or aboard ships. They typically use computers to perform their work. These jobs can sometimes be routine and monotonous, and concentration and attention to detail are critical.

OUTLOOK

Employment in the armed forces is expected to grow about as fast as the average for all occupations through 2014, according to the U.S. Department of Labor. When the economy is good and/or during times of war, more people pursue employment in the civilian workforce, which creates additional opportunities in the military. With the U.S. military involved in several international conflicts as this book went to press, most significantly in Iraq and Afghanistan, demand should continue to be strong for military workers.

Opportunities for military workers in accounting, budget, and finance should also be good. They are responsible for managing billions of dollars, and the military would not be able to function without their services.

The employment outlook in the civilian sector varies by occupation. The U.S. Department of Labor predicts that employment for accountants will grow faster than the average for all occupations through 2014. Employment for bookkeeping and accounting clerks is expected to grow more slowly than the average.

FOR MORE INFORMATION

For information on career opportunities, contact

American Institute of Professional Bookkeepers
6001 Montrose Road, Suite 500

Rockville, MD 20852-4873
Tel: 800-622-0121
Email: info@aipb.org
http://www.aipb.org

For information on accredited programs in accounting, contact
Association to Advance Collegiate Schools of Business
777 South Harbour Island Boulevard, Suite 750
Tampa, FL 33602-5730
Tel: 813-769-6500
http://www.aacsb.edu

For more information on women in accounting, contact
The Educational Foundation for Women in Accounting
PO Box 1925
Southeastern, PA 19399-1925
Tel: 610-407-9229
Email: info@efwa.org
http://www.efwa.org

For information about management accounting, as well as student membership, contact
Institute of Management Accountants
10 Paragon Drive
Montvale, NJ 07645-1718
Tel: 800-638-4427
Email: ima@imanet.org
http://www.imanet.org

To get information on specific branches of the military, check out this site, which is the home of ArmyTimes.com, NavyTimes.com, AirForceTimes.com, and MarineCorpsTimes.com:
Military City
http://www.militarytimes.com

If you're thinking of joining the armed forces, take a look at this site, which guides students and parents through the decision-making process:
Today's Military
http://www.todaysmilitary.com

For information on military careers, contact
United States Air Force
http://www.airforce.com

United States Army
http://www.goarmy.com

United States Coast Guard
http://www.gocoastguard.com

United States Marine Corps
http://www.marines.com

United States Navy
http://www.navy.com

Art, Design, and Music Occupations

QUICK FACTS

School Subjects
Art
Computer science
Music

Personal Skills
Artistic
Communication/ideas

Work Environment
Primarily indoors
Primarily multiple locations

Minimum Education Level
Varies by career specialty

Salary Range
$15,617 to $30,618 to $66,154 (enlisted personnel)
$25,358 to $70,877 to $174,103 (officers)

Outlook
About as fast as the average

DOT
141, 143, 152, 651

GOE
01.04.01, 01.04.02, 01.05.02, 01.07.01, 01.08.01, 08.03.05

O*NET-SOC
11-2031.00, 27-1013.00, 27-1013.01, 27-1024.00, 27-2042.00, 27-2042.02, 27-3031.00, 27-4021.02, 51-5023.09

OVERVIEW

Military workers in art and design occupations design, produce, and present information and ideas. Others give live musical performances. Opportunities for both enlisted personnel and officers are available in the U.S. Air Force, Army, Coast Guard, Marines, and Navy.

HISTORY

The visual arts and music were essential elements of many early cultures and civilizations. Printing was not invented until 868 A.D. in China. It was not until 1827 that the first photograph was produced in France. Today, these disciplines and others fueled by the development of computers and the Internet provide us with a variety of creative and noncreative resources such as newspapers, books, magazines, paintings, illustration, computer art, photographs, music, Web sites, and computer and video games.

Art and design professionals have played an important role in the U.S. military for many years. Military musicians and photographers have some of the most interesting history.

In the early 1700s, a military band of wind instrumentalists performed at Faneuil Hall in Boston; this is considered by many to be the first organized performance by a military band in the American colonies. In the events leading up to the Revolutionary War, musicians helped minutemen drill. During the war, some regiments were known to have bands, mainly drummers and trumpeters. During the Civil War, Congress passed an act, according to the U.S. Army Bands Web site, that "authorized each Regular Army regiment of infantry two principal musicians per company and 24 musicians for a band." Musicians were also authorized for artillery and cavalry regiments, with each artillery band permitted 24 musicians and each cavalry band permitted 16 musicians. Military musicians played important roles in peacetime and in all subsequent wars, including World Wars I and II, the Vietnam War, the Gulf War, and the wars in Afghanistan and Iraq. In fact, eight military bands served in the Gulf War, and, today, musicians entertain troops in Iraq as well as those stationed in the United States. Military bands also perform ceremonial duties and entertain the public at concert halls and other venues.

An Englishman named Roger Fenton was the first person to take photographs on a battlefield (during the Crimean War, 1854–56) while fighting raged. Matthew Brady took photographs of battlefields during the Civil War, but these were mostly taken for civilians and rarely caught scenes of battle. The Union army attempted to take aerial photos of enemy encampments in 1862, according to *Overview: A Life-Long Adventure in Aerial Photography* (Doubleday, 1969), but did not have much success due to technical limitations. It was not until World War I that photography was successfully used for strategic purposes. In 1915, the British used aerial photographs to create maps of enemy trenches. By the end of the war, both sides began using photography to gather intelligence, as well as for training and propaganda. Today, aerial photography from satellites is a key tool used in intelligence gathering. In addition to use in intelligence, photographs are used in public relations, journalism, and for training purposes.

THE JOB

Some of the more popular options for those interested in art and design careers in the military are described below.

Graphic designers and illustrators design or illustrate a wide variety of graphic artwork used by the military, including graphs, charts, art, and computer graphics. These items are used in the military's training and promotional materials, publications, visual displays and presentations, signage, or scenery for film and television broadcasts.

Military bands often perform at parades, concerts, and presentations. *Music directors* plan performances and direct military bands during official engagements. Their duties include overseeing the training and practice of musicians and choir members, designing special musical programs, and conducting the band during performances. They also create band and choir budgets and initiate purchases for instruments and equipment.

In addition to performing music, *musicians* in the military also spend time practicing, composing, and arranging music. They may work alone or as part of a band and generally play before live audiences outdoors, in clubs, or in auditoriums. They typically perform at ceremonies, festivals, or parades. Musicians in the military play many different types of music such as classical, popular (including rock and country), jazz, and traditional military marches.

Photographic specialists take, develop, and print pictures, using a variety of cameras and photographic equipment. They take pictures of people, places, objects, and events to aid with such operations as news reporting, promotional activities, or gathering military intelligence. They develop the photographs and create slides, negatives, or duplicate copies. They also use digital cameras. Photographic specialists also maintain the equipment they use.

Printing specialists prepare, operate, and maintain printing presses used to produce maps, newspapers, pamphlets, charts, training materials, and miscellaneous publications used by the military. Their duties include installing and adjusting printing plates, loading and feeding paper, mixing inks and controlling ink flow, and then binding the printed pages together to produce the finished product, ensuring the quality of the final printed piece.

REQUIREMENTS

High School

While in high school, take as many art, photography, and graphic design classes as are available. It is also a good idea to take English, speech, and drama classes.

If you are interested in music, take band, orchestra, or choir classes depending on your interest. In addition, you should also take mathematics classes, since any musician needs to understand counting, rhythms, and beats. Many professional musicians write at least some of their own music, and a strong math background is very helpful for this. If your high school offers courses in music history or appreciation, be sure to take these as well.

If you are interested in becoming a printing specialist, you should take courses that develop your mechanical and mathematical aptitude, such as shop, mathematics, and computer science.

Postsecondary Training

Educational requirements in the civilian sector vary by career. For example, graphic designers and illustrators typically have some postsecondary training; a growing number have bachelor's degrees in commercial art or fine art. Music professionals usually attend two- or four-year colleges to hone their musical talents. A college educa-

Rank and Military Branch by Occupation

Job Title	Rank	Military Branches
Graphic Designers and Illustrators	Enlisted	Air Force, Army, Marines, Navy
Music Directors	Officer	Air Force, Army, Coast Guard, Marines, Navy
Musicians	Enlisted	Air Force, Army, Coast Guard, Marines, Navy
Photographic Specialists	Enlisted	Air Force, Army, Coast Guard, Marines, Navy
Printing Specialists	Enlisted	Army, Marines, Navy

Source: U.S. Department of Defense

tion is not required to become a photographer, although college training probably offers the most promising assurance of success in fields such as industrial, news, or scientific photography. Printing professionals train for the field via apprenticeships or through post-secondary programs in printing equipment operation.

Military workers in this specialty receive job training in classroom settings, on the job, and through advanced courses. Course work varies by specialty. For example, photographic specialists learn the principles of photojournalism; how to take, process, and reproduce photos; and how to operate and maintain cameras, scanners, and other photographic equipment. Musicians, on the other hand, must be musically proficient before they enter the military. Once they join the military they continue to hone their talents by taking courses on music theory, group instrumental techniques, and other topics, as well as participating in regular rehearsals and individual practice. Music directors need to have a bachelor's degree before they join the military.

Visit the U.S. Department of Defense's Web site, http://www.todaysmilitary.com, for more on military training for those interested in art and design careers.

Other Requirements

Personal requirements for workers in these occupations vary by specialty. For example, graphic designers, illustrators, and photographic specialists should have artistic ability. Musicians must have musical talent and the ability to perform in public. Music directors should have strong organizational skills and enjoy all types of music. Printing specialists must have detailed knowledge of printing processes and be willing to do work that is sometimes physically demanding.

The Air Force Band plays ceremonial music during the services honoring the life and legacy of former President Gerald R. Ford as he is laid to rest in Grand Rapids, Michigan. *(Staff Sergeant Helen Mill, U.S. Army, U.S. Department of Defense)*

Visit the U.S. Department of Defense's Web site, http://www.todaysmilitary.com, for more on personal requirements for workers in these careers.

EXPLORING

If you are interested in a career in art and design, you can find out whether you have the talent, ambition, and perseverance to succeed in these fields in a number of ways. Take as many art, design, illustration, and photography courses as possible while still in high school and become proficient at working on computers. To get an insider's view of these various occupations, enlist the help of teachers or school guidance counselors to make arrangements to tour design firms, art studios, or photo studios and interview professionals in the field.

The first step to exploring your interest in a musical career is to become involved with music. Elementary schools, high schools, and institutions of higher education all present a number of options for musical training and performance, including choirs, ensembles, bands, and orchestras. You also may have chances to perform in school musicals and talent shows. Those involved with services at churches, synagogues, or other religious institutions have excellent opportunities to explore their interest in music. If you can afford to, take private music lessons. You may also want to attend special summer camps or programs that focus on the field.

Others ways to learn about any of the jobs listed in this article include reading books and magazines about these careers, visiting Web sites of professional associations, contacting typical employers to arrange a tour of their facilities (such as photography or graphic design studios), and asking your guidance counselor or teacher to arrange an information interview with a worker in the field.

To learn more about career opportunities in the military, visit the Web sites listed at the end of this article.

EMPLOYERS

The U.S. government employs the military. No specific employment statistics are available for military workers in art and design occupations. Overall, 1.4 million men and women are on active duty and another 1.2 million volunteers serve in the National Guard and Reserve forces.

STARTING OUT

If you are interested in entering the armed forces, you should first contact a military recruiter. Visit the Web sites listed at the end of this article to locate a recruiting office near you. To start out in any branch, you will need to pass physical and medical tests, the Armed Services Vocational Aptitude Battery exam, and basic training.

ADVANCEMENT

Each military branch has nine enlisted grades (E-1 through E-9) and 10 officer grades (O-1 through O-10). The higher the number is, the more advanced a person's rank is. The various branches of the military have somewhat different criteria for promoting individuals; in general, however, promotions depend on factors such as length of time served, demonstrated abilities, recommendations, and scores on written exams. Promotions become more and more competitive as people advance in rank. On average, a diligent enlisted person can expect to earn one of the middle noncommissioned or petty officer rankings (E-4 through E-6); some officers can expect to reach lieutenant colonel or commander (O-5). Outstanding individuals may be able to advance beyond these levels.

EARNINGS

The U.S. Congress sets the pay scales for the military after hearing recommendations from the president. The pay for equivalent grades is the same in all services (that is, anyone with a grade of E-4, for example, will have the same basic pay whether in the army, navy, marines, air force, or Coast Guard). In addition to basic pay, personnel who frequently and regularly participate in combat may earn hazardous duty pay. Other special allowances include special duty pay and foreign duty pay. Earnings start relatively low but increase on a fairly regular basis as individuals advance in rank. See the appendix at the end of this book for detailed information on pay scales for the U.S. military. When reviewing earnings, it is important to keep in mind that members of the military receive free housing, food, and health care—items that civilians typically pay for themselves.

Additional benefits for military personnel include uniform allowances, 30 days' paid vacation time per year, and the opportunity to retire after 20 years of service. Generally, those retiring will receive 40 percent of the average of the highest three years of their base pay. This amount rises incrementally, reaching 75 percent of the average of the highest three years of base pay after 30 years of service. All retirement provisions are subject to change, however, and you should verify them as well as current salary information before you enlist. Those who retire after 20 years of service are usually in their 40s and thus have plenty of time, as well as an accumulation of skills, with which to start a second career.

WORK ENVIRONMENT

Those in art and design occupations typically work in offices or studios. Music directors, musicians, and photographic specialists sometimes work outdoors in inclement weather. Photographic specialists may take photos from ships or aircraft.

OUTLOOK

Employment in the armed forces is expected to grow about as fast as the average for all occupations through 2014, according to the U.S. Department of Labor. When the economy is good and/or during times of war, more people pursue employment in the civilian workforce, which creates additional opportunities in the military. With the U.S. military involved in several international conflicts as this book went to press, most significantly in Iraq and Afghanistan, demand should continue to be strong for military workers. Military opportunities should also be good for workers in this small, but important, career field.

The employment outlook in the civilian sector varies by occupation. The U.S. Department of Labor offers the following predictions for professions in the field: graphic designers and illustrators, music directors, musicians, and photographic specialists—about as fast as the average; printing specialists—more slowly than the average.

FOR MORE INFORMATION

For information on membership in a local union near you, developments in the music field, a searchable database of U.S. and foreign music schools, and articles on careers in music, contact

American Federation of Musicians of the United States and Canada
1501 Broadway, Suite 600
New York, NY 10036-5505
Tel: 212-869-1330
http://www.afm.org

For more information about careers in graphic design, contact

American Institute of Graphic Arts
164 Fifth Avenue
New York, NY 10010-5901
Tel: 212-807-1990
http://www.aiga.org

This graphic arts trade association is a good source of general information about the printing industry.

National Association for Printing Leadership
75 West Century Road
Paramus, NJ 07652-1408
Tel: 201-634-9600
Email: Information@napl.org
http://www.napl.org

This organization provides training, publishes its own magazine, and offers various services for its members.

Professional Photographers of America
229 Peachtree Street, NE, Suite 2200
Atlanta, GA 30303-1608
Tel: 800-786-6277
Email: csc@ppa.com
http://www.ppa.com

This organization promotes and stimulates interest in the art of illustration by offering exhibits, lectures, educational programs, and social exchange. For information, contact

Society of Illustrators
128 East 63rd Street
New York, NY 10021-7303
Tel: 212-838-2560

Email: info@societyillustrators.org
http://www.societyillustrators.org

To get information on specific branches of the military, check out this site, which is the home of ArmyTimes.com, NavyTimes.com, AirForceTimes.com, and MarineCorpsTimes.com:

Military City
http://www.militarytimes.com

If you're thinking of joining the armed forces, take a look at this site, which guides students and parents through the decision-making process:

Today's Military
http://www.todaysmilitary.com

For information on military careers, contact

United States Air Force
http://www.airforce.com

United States Army
http://www.goarmy.com

United States Coast Guard
http://www.gocoastguard.com

United States Marine Corps
http://www.marines.com

United States Navy
http://www.navy.com

For information on musical careers in the armed forces, visit

United States Air Force Band
http://www.usafband.af.mil

United States Army Band
http://www.usarmyband.com
http://bands.army.mil

United States Coast Guard Band
http://www.uscg.mil/band

United States Marine Band
http://www.marineband.usmc.mil

United States Navy Band
http://www.navyband.navy.mil

INTERVIEW

Mark Weaver is the senior chief musician, principal trombonist, for the U.S. Coast Guard Band (http://www.uscg.mil/band). He has served in the military for 28 years. Mark discussed his career with the editors of Careers in Focus: Armed Forces.

Q. Take us through a day in your life as a Coast Guard musician. What are your typical tasks/responsibilities?

A. First and foremost are my duties as the trombone section leader. I assign members of the trombone section to jobs that come up during our fiscal year. Those jobs can range from ceremonies to small group (brass quintet, ceremonial band) performances mainly in Washington, D.C.

An average day begins with a muster (or meeting of all band members), where word is passed in regard to the day's activities. Rehearsals begin at 9:00 A.M. and go until 11:30 with a 20-minute break in the middle. Afternoons are used for administrative work. Each member of the Coast Guard Band has a collateral duty that can range from load crew to graphic artist. In my case, I am the band's graphic artist. I produce all of the band's color computer-generated artwork for CDs, brochures, programs, fliers, etc. In addition, as a senior member of the band, I serve on the band's staff. The staff is charged with taking care of all of the issues directed to the band through our chain of command, which includes support of the band director's initiatives as established in our mission statement.

Q. What is your work environment like?

A. We have a pretty good auditorium to rehearse and perform in. Leamy Hall, which is located at the U.S. Coast Guard Academy in New London, Conn., was built in the 60's and is now getting a little dated. We do all of our recording there as well. It seats around 1,500 people, which we usually fill for our monthly concert series. Our office area is in the basement of Leamy Hall. It's not the greatest.

Q. Do you travel for your job?

A. Yes, about an average of 40 days total per year. That includes ceremonial trips to Washington, D.C., and a national concert tour, which is about 14 to 20 days in length. On occasion, the band travels overseas. During my tenure I have performed in

Russia, the United Kingdom, and in 47 out of the 50 states. This year (2008) the band will travel to Japan.

Q. What do you like most about your job?

A. I like the travel, and I really enjoy working with my colleagues in the trombone section. It's one of the best trombone sections in military music. There are personal things that I like as well, such as ceremonies at Arlington National Cemetery. Playing in the cemetery gives me a feeling of making a real contribution to our service and country. Honoring those who have served and fallen in the line of duty is a very special thing.

Q. What are the three most important professional qualities for military musicians?

A. Professionalism, respect, and devotion to duty.

Q. What advice would you give to high school students who are interested in becoming musicians?

A. Practice the fundamentals. Work to develop the best sound you can make. It doesn't matter how fast or how many notes you can play if you don't have a great sound. By working on the basics, tackling repertoire will make more sense. There are no shortcuts in the music world. Only hard work and careful study will get you a job playing. I can't tell you how many auditions were lost by players who either could not play in time, or just sounded like they were on the edge. You can really tell who has been working on things in a methodic manner.

Q. What advice would you give to those entering the military?

A. Always remember you are entering the service. "Service" is the operative word. You are there to serve others. You must learn to work under the system of a chain of command. You must be ready to take orders before you can give them. A lot of people forget this, and it can be a real roadblock for them during their career. On the other hand, there can be a great return. I can't tell you how much people appreciate our performances. We represent our service in a way that no other unit in the Coast Guard can. There have been many proud moments during my career.

Aviation Occupations, Except Pilots

OVERVIEW

Many different types of aircraft are deployed by the military for use in missions including combat, search and rescue, transportation of troops and equipment, and air-dropping cargo. *Aviation professionals* keep these aircraft in top mechanical condition; direct them during takeoff, in-flight, and when landing; and provide many other support duties to ensure the success of military aviation operations. (See this book's article "Military Pilots" for information on opportunities for pilots in the military. Pilots serve in the U.S. Air Force, Army, Coast Guard, Marines, and Navy.)

QUICK FACTS

School Subjects
Computer science
Technical/shop

Personal Skills
Mechanical/manipulative
Technical/scientific

Work Environment
Indoors and outdoors
Primarily multiple locations

Minimum Education Level
Varies by career

Salary Range
$15,617 to $30,618 to $66,154 (enlisted personnel)
$25,358 to $70,877 to $174,103 (officers)

Outlook
About as fast as the average

DOT
193, 621

GOE
05.03.01, 07.02.01

O*NET-SOC
49-2091.00, 49-3011.00, 49-3011.01, 49-3011.02, 49-3011.03, 53-2021.00, 53-2022.00

HISTORY

The age of modern aviation is generally considered to have started when Orville and Wilbur Wright successfully flew their heavier-than-air machine on December 17, 1903. The government soon realized the great possibilities for aviation, and created the Army Aeronautical Division in 1907. Technological advances followed rapidly. Spurred in part by the growing certainty of war in Europe, designers sought faster, more maneuverable, and more stable aircraft. By the end of World War II, planes could reach speeds nearing 200 miles per hour and could climb as high as 30,000 feet, while operating for many hours in the air.

Air power played a major role in World War II. The Allied powers, mainly the United States and Great Britain, had strong air forces that were able to eventually wrest control of enemy airspace from the Axis powers (namely Germany and Japan) and turn the tide of the war. The U.S. also used its overwhelming air power to win the war in the Pacific against Japan once the Germans were defeated. It dropped nuclear bombs on the Japanese cities of Hiroshima and Nagasaki—forcing Japan to surrender. Recognition of the strategic importance of air power led to the creation of a new branch of the armed forces, the U.S. Air Force, in 1947.

From the early days of military aviation, through World War II, to today's wars in Afghanistan and Iraq, aviation professionals have played a key role in keeping aircraft taking off, flying, and landing safely.

THE JOB

Air traffic controllers monitor and direct the activities of aircraft into and out of military airfields and along specified flight routes. They radio pilots with approach, landing, taxiing, and takeoff instructions and advisories on weather and other conditions to maintain the safe and orderly flow of air traffic both in the air and on the ground.

Air traffic controller managers supervise air traffic controllers and other personnel who work in various sections of air traffic control centers. These include those who direct pilots during takeoffs and landing, those who provide ground instructions, and those who monitor planes during flight.

Aircraft launch and recovery specialists operate various equipment used for military launch and recovery missions. Their duties include directing aircraft launch and recovery operations and testing and operating equipment such as catapult systems used to launch aircrafts from aircraft carriers and temporary airfields. They also install, maintain, and repair landing aids such as arresting gear and barricades.

Aircraft mechanics examine, service, repair, and overhaul aircraft and helicopters and their engines. They also repair, replace, and assemble parts of the airframe (the structural parts of the plane other than the power plant or engine).

Flight engineers inspect and monitor the performance of military aircrafts to ensure they are safe, efficient, and reliable. Before, during, and after a flight, engineers check the engines, equipment, and systems of an airplane or helicopter. They may tweak systems, such as adjusting the electrical boost pump to make better use of fuel. Their duties include helping pilots during engine start-up and shutdown, working with an airplane's weight load capabilities, or fuel distribution. Flight engineers often confer with pilots regarding problems with their aircrafts, and make suggestions on how to correct them.

Flight navigators use radar, radio, and other equipment to help military pilots determine aircraft position and determine its route of travel. Typical duties include guiding aircraft during in-flight refueling operations, providing data (such as altitude, fuel usage, etc.) to pilots during flight, helping pilots navigate the aircraft during flight, operating radios to send and receive communications, and testing navigation and weapons systems before flights.

Flight operations specialists prepare and record flight information on the military's many passenger, transport, and combat aircrafts. They maintain logs for incoming and outgoing flights, keep records for air crew, and plan schedules for air flights and crew assignments. They also advise pilots and crew members regarding flight plans and weather reports. Some specialists are responsible for an airplane's evacuation equipment or other necessary items.

REQUIREMENTS

High School

If you are interested in becoming an aircraft mechanic, take machine shop, auto mechanics, and electrical shop. Courses in mathematics, physics, chemistry, and mechanical drawing are particularly helpful because they teach the principles involved in the operation of an aircraft, and this knowledge is often necessary to make the repairs.

Rank and Military Branch by Occupation

Job Title	Rank	Military Branches
Air Crew Members	Enlisted	Air Force, Army, Coast Guard, Marines, Navy
Air Traffic Control Managers	Officer	Air Force, Army, Marines, Navy
Air Traffic Controllers	Enlisted	Air Force, Army, Marines, Navy
Aircraft Launch and Recovery Specialists	Enlisted	Air Force, Army, Coast Guard, Navy
Airplane Navigators	Officer	Air Force, Army, Coast Guard, Marines, Navy
Flight Engineers	Enlisted	Air Force, Marines, Navy
Flight Operations Specialists	Enlisted	Air Force, Army, Coast Guard, Marines, Navy

Source: U.S. Department of Defense

If you are interested in becoming an air traffic controller, you should pursue a college prep curriculum. Mathematics and science courses are especially useful courses to study because they are most directly related to air traffic control work.

Postsecondary Training

Educational requirements in the civilian sector vary by career. Aircraft mechanics typically earn a one-year certificate or two-year associate degree from a postsecondary training institution that has been approved by the Federal Aviation Administration (FAA). Air traffic controllers must complete training at a facility that is approved by the FAA. They must also pass a preemployment test that gauges their aptitude for the career and have completed four years of college or have three years of work experience or a combination of both.

In the military, enlisted personnel (which include every occupation listed in this article except air traffic control managers and airplane navigators) train via classroom instruction, on-the-job experience, and advanced course work. Course work varies by specialty. For example, air traffic controllers learn how to do their jobs by taking classes in air traffic control management, communications and radar procedures, and takeoff, landing, and ground control procedures. Aircraft mechanics take classes that teach them how to repair engines and disassemble and repair aircraft systems, airframes, and coverings.

As officers, air traffic control managers must enter the military with at least a bachelor's degree. Once they join the military, they take advanced versions of the classes that are taken by air traffic controllers, as well as classes on management. Both air traffic controllers and air traffic controller managers who work on aircraft carriers receive specialized training to prepare them for this demanding specialty. Airplane navigators learn how to do their jobs via classroom instruction, training in aircraft simulators, and through approximately 100 hours of actual flying time.

Visit the U.S. Department of Defense's Web site, http://www.todaysmilitary.com, for more on military training in aviation occupations.

Other Requirements

Personal requirements for workers in these occupations vary by specialty. For example, air traffic controllers and managers should be able to express themselves clearly, remember rapidly changing data that affect their decisions, and be able to operate calmly under very difficult situations involving a great deal of strain. They must also be able to make good, sound, and quickly derived decisions. Aircraft launch and recovery specialists and aircraft mechanics must be able to work with precision and meet rigid standards. Their physical condition is also important. They need more than average strength for lifting heavy parts and tools, as well as agility for reaching and climbing. And they should not be afraid of heights, since they may work on top of the wings and fuselages of large jet planes.

Visit the U.S. Department of Defense's Web site, http://www.todaysmilitary.com, for more on personal requirements for workers in these careers.

EXPLORING

Working with electronic kits and assembling model airplanes are good ways of gauging your ability and seeing if you are interested in working in one of the more mechanical-oriented careers in this specialty. Kits for building ultralight craft are also available and may provide even more insight into the importance of proper maintenance and repair.

Others ways to learn about any of the jobs listed in this article include reading books and magazines about these careers, visiting Web sites of professional associations, taking a tour of an airfield or air traffic control tower, and asking your guidance counselor or

Active Duty Officers by Rank

Rank/Grade	Total (Air Force, Army, Marine Corps, Navy)
Generals/Admirals	40
Lt. Generals/Vice Admirals	132
Major Generals/Rear Admirals	283
Brigadier Generals	447
Colonels	11,296
Lieutenant Colonels	27,610
Majors	44,116
Captains	70,057
1st Lieutenants	23,214
2nd Lieutenants	26,306
Chief Warrant Officers	17,407
Total Officers	**220,908**

Source: U.S. Department of Defense, October 31, 2007

teacher to arrange an information interview with a worker in the field, such as an air traffic controller or aircraft mechanic.

To learn more about career opportunities in the military, visit the Web sites listed at the end of this article.

EMPLOYERS

The U.S. government employs the military. No specific employment statistics are available for aviation workers in the military. Overall, 1.4 million men and women are on active duty and another 1.2 million volunteers serve in the National Guard and Reserve forces.

STARTING OUT

Contact a military recruiter for more information about the variety of jobs available in the armed forces. Visit the Web sites listed at the end of this article to locate a recruiting office near you. To start out in any branch, you will need to pass physical and medical tests, the Armed Services Vocational Aptitude Battery exam, and basic training.

ADVANCEMENT

Each military branch has nine enlisted grades (E-1 through E-9) and 10 officer grades (O-1 through O-10). The higher the number is, the more advanced a person's rank is. The various branches of the military have somewhat different criteria for promoting individuals; in general, however, promotions depend on factors such as length of time served, demonstrated abilities, recommendations, and scores on written exams. Promotions become more and more competitive as people advance in rank. On average, a diligent enlisted person can expect to earn one of the middle noncommissioned or petty officer rankings (E-4 through E-6); some officers can expect to reach lieutenant colonel or commander (O-5). Outstanding individuals may be able to advance beyond these levels.

EARNINGS

The U.S. Congress sets the pay scales for the military after hearing recommendations from the president. The pay for equivalent grades is the same in all services (that is, anyone with a grade of E-4, for example, will have the same basic pay whether in the army, navy, marines, air force, or Coast Guard). In addition to basic pay, personnel who frequently and regularly participate in combat may earn hazardous duty pay. Other special allowances include special duty pay and foreign duty pay. Earnings start relatively low but increase on a fairly regular basis as individuals advance in rank. See the appendix at the end of this book for detailed information on pay scales for the U.S. military. When reviewing earnings, it is important to keep in mind that members of the military receive free housing, food, and health care—items that civilians typically pay for themselves.

Additional benefits for military personnel include uniform allowances, 30 days' paid vacation time per year, and the opportunity to retire after 20 years of service. Generally, those retiring will receive 40 percent of the average of the highest three years of their base pay. This amount rises incrementally, reaching 75 percent of the average of the highest three years of base pay after 30 years of service. All retirement provisions are subject to change, however, and you should verify them as well as current salary information before you enlist. Those who retire after 20 years of service are usually in their 40s and thus have plenty of time, as well as an accumulation of skills, with which to start a second career.

WORK ENVIRONMENT

The work environment for aviation workers depends a great deal on their specialty. For example, air traffic controllers and managers work in clean, climate-controlled air traffic control centers at military bases and on ships. Aircraft launch and recovery specialists often work outdoors on ships in all types of weather while operating and maintaining launch and recovery equipment or providing instruction to aircraft to help them land. They are also exposed to loud noise and fumes from jet and helicopter engines. Aircraft mechanics work in machine shops and aircraft hangars on air bases or aboard aircraft carriers. When conducting major overhauls and inspections, they generally work in hangars with adequate ventilation, lighting, and heat. If the hangars are full, however, or if repairs must be made quickly, they may work outdoors, sometimes in extreme heat or cold or other demanding weather conditions.

OUTLOOK

Employment in the armed forces is expected to grow about as fast as the average for all occupations through 2014, according to the U.S. Department of Labor. When the economy is good and/or during times of war, more people pursue employment in the civilian workforce, which creates additional opportunities in the military. With the U.S. military currently involved in several international conflicts, most significantly in Iraq and Afghanistan, demand should continue to be strong for military workers.

Opportunities should be very good for military aviation workers. To retain its supremacy in the air, the U.S. military will continue to need qualified workers to keep aircraft flying safely and effectively.

Employment for aviation workers in the civilian sector is expected to grow about as fast as the average for all occupations through 2014, according to the U.S. Department of Labor.

FOR MORE INFORMATION

For information on air traffic control and scholarships, contact

Air Traffic Control Association
1101 King Street, Suite 300
Alexandria, VA 22314-2963
Tel: 703-299-2430
Email: info@atca.org
http://www.atca.org

For career information, contact

Federal Aviation Administration
800 Independence Avenue, SW
Washington, DC 20591-0001
Tel: 202-366-4000
http://www.faa.gov

For information on aviation maintenance and scholarships, contact

Professional Aviation Maintenance Association
400 Commonwealth Drive
Warrendale, PA 15096-0001
Tel: 866-865-7262
Email: hq@pama.org
http://www.pama.org

To get information on specific branches of the military, check out this site, which is the home of ArmyTimes.com, NavyTimes.com, AirForceTimes.com, and MarineCorpsTimes.com:

Military Times
http://www.militarytimes.com

If you're thinking of joining the armed forces, take a look at this site, which guides students and parents through the decision-making process:

Today's Military
http://www.todaysmilitary.com

For information on military careers, contact

United States Air Force
http://www.airforce.com

United States Army
http://www.goarmy.com

United States Coast Guard
http://www.gocoastguard.com

United States Marine Corps
http://www.marines.com

United States Navy
http://www.navy.com

Combat Specialty Occupations

QUICK FACTS

School Subjects
Government
Physical education

Personal Skills
Following instructions
Leadership/management

Work Environment
Indoors and outdoors
Primarily multiple locations

Minimum Education Level
Varies by career specialty

Salary Range
$15,617 to $30,618 to $66,154 (enlisted personnel)
$25,358 to $70,877 to $174,103 (officers)

Outlook
About as fast as the average

DOT
378

GOE
04.05.01, 04.05.02

O*NET-SOC
55-1015.00, 55-1019.99, 55-3019.99

OVERVIEW

Combat specialty workers operate various weapons systems and perform special offensive or defensive missions. Units specialize according to the type of weapons system or combat operation, ranging from long-range missiles and armored assault vehicles, to the search and rescue missions of the elite Special Forces. The U.S. Air Force, Army, Marines, Navy, and Coast Guard offer opportunities for both enlisted personnel and officers. There were 207,574 soldiers employed in combat specialty occupations in 2005.

HISTORY

The history of the U.S. military dates back to defense forces, known as militias, that were used by the colonies. These militias began to develop in the first decades of the 17th century, long before the United States existed as a country. More than 100 years later, in 1775, the Continental Army was established to fight the British in the Revolutionary War. The colonists so valued the army that its commander and most revered general, George Washington, became the first president of the United States.

The oldest continuous seagoing service in the United States, the Coast Guard, was established in 1790 to combat smuggling. In contrast, the first American marine units were attached to the army at the time of its creation; these units then were made an independent part of the navy when it was officially established in 1798. The Marine Corps was considered part of the navy until 1834, when it established itself as both a land and sea defense force, thereby becoming its own military branch.

The air service grew from somewhat unusual beginnings. The Civil War marked the first use of aircraft in the U.S. military, when a balloon corps was attached to the Army of the Potomac. In 1892, a formal Balloon Corps was created as part of the army's Signal Corps. By 1907, a separate aeronautical division was created within the army. Air power proved invaluable a few years later during World War I, bringing about major changes in military strategy. As a result, the United States began to assert itself as an international military power, and accordingly, the Army Air Service was created as an independent unit in 1918, although it remained under army direction for a time.

With the surprise attack on Pearl Harbor in 1941, America was plunged into World War II. At its height, 13 million Americans fought in the different branches of the military services. When the war ended, the United States emerged as the strongest military power in the Western world. A large part of America's military success was due to the superiority of its air forces. Recognition of the strategic importance of air power led to the creation of the now wholly independent branch of service, the U.S. Air Force, in 1947. Two years later, the various branches of military service were unified under the Department of Defense. (Today, the U.S. Coast Guard falls under the jurisdiction of the U.S. Department of Homeland Security.)

In the years following World War II, the United States and its allies devoted considerable military resources to fighting the Cold War with the Soviet Union. Anticommunist tensions led to U.S. involvement in the Korean War during the 1950s and to participation in the Vietnam War, which ended in the mid-1970s. Antiwar sentiment grew increasingly insistent, and soon, the policies that established an American presence in foreign countries came under new demands for re-evaluation. In 1973, the draft was abolished, and the U.S. military became an all-volunteer force. The armed forces began to put great energy into improving the image of military personnel and presenting the military as an appealing career option to attract talented recruits.

During the 1980s, the U.S. military increased its efforts to bring about the collapse of Soviet communism and became active in the Middle East, particularly in the Persian Gulf, through which flowed much of the world's oil supply. Later in the decade, many countries under Soviet rule began to press for independence, and, in 1991, the Soviet Union finally collapsed under the weight of its political and economic crisis, effectively ending the Cold War. That same year, the United States engaged in the Persian Gulf War.

From the early 1990s up until the terrorist attacks of September 11, 2001, the U.S. military took on a new role as a peacekeeping force. It participated in cooperative efforts led by organizations such as the United Nations and the North Atlantic Treaty Organization.

Reaction to the terrorist attacks on the United States suddenly changed the role of the military from a peacekeeping force to an aggressor in the attempt to destroy the strongholds and training camps of terrorists around the world. President George W. Bush said the war against terrorism would likely be a sustained effort over a long period of time.

THE JOB

During peacetime, all combat specialty workers stand ready to defend the United States. They practice methods of warfare, learn how to operate new equipment, and otherwise prepare themselves in the event they are called to defend the country. Specialties in the field are described below.

Armored assault vehicle crew members operate combat vehicles such as tanks and armored land or amphibious assault vehicles, and engage artillery weapons against the enemy. Some crew members may be trained to gather intelligence regarding the enemy's location and strength and give their reports to officers. Other important tasks include maintaining vehicles and weapons, operating signaling equipment, and interpreting maps and battle plans.

Armored assault vehicle officers lead their crew members in military combat formations using different units that specialize in the operation of amphibious assault, armored tanks, and cavalry vehicles. In addition to assault vehicles, they also supervise the use and maintenance of support equipment such as weapons and computerized positioning and communications equipment. Officers also formulate battle plans, communicate with other unit officers, and oversee the strategic and technical training of their unit members.

The power of artillery and missiles is imperative during wartime. *Artillery and missile crew members* are assigned to protect troops from enemy assault on land and sea using different weapon systems such as cannons, howitzers, missiles, or rockets. They may also launch these weapons by land, sea, or air in order to destroy the enemy's bases, troops, aircraft, and ships. Each crew, specializing in a particular type of system, is responsible for the preparation of ammunition, and the operation and maintenance of their weapons. Crew members are also trained to use computerized equipment to locate and identify enemy targets.

Artillery and missile officers are assigned to a crew for training, direction, and supervision. They are responsible for ensuring that every crew member knows how to use and maintain their weapons and related equipment. Some officers may also keep track of artillery supplies, including major weaponry such as tanks, missiles, and cannons. Officers often work with other combat units to identify and select targets for artillery and missile strikes. Officers specialize in a particular system as well as location—for example, *field artillery officers* may be assigned to fight and suppress enemies on the battlefield using cannons, rockets, and missile power.

Combat mission support officers plan and implement battle strategies. In addition to providing battle management support, they train and evaluate personnel within their units. Officers may also advise unit commanders, according to their area of expertise, on possible deployment of forces or other actions. For example, an air force electronics combat officer may be responsible for interpreting radar signals to identify incoming enemy aircrafts or missiles.

Infantry workers provide the main land combat force during wartime. They arrive on the battlefield armed with weapons and supplies in order to attack enemy forces or repel an attack. Infantry workers establish camps and locations by digging foxholes, trenches, and bunkers, or by setting up camouflaged and protected barriers. Some infantry workers may be sent on scouting missions to identify the enemy's troops and ammunition locations. They may resort to hand-to-hand combat in order to capture or destroy enemy forces. Only men are allowed to serve in the infantry.

Infantry officers train and lead units of infantry workers. They call for the use of weapons including machine guns, rocket launchers, mortars, and tanks and armored personnel carriers. Using information they've gathered on the size and position of the enemy, infantry officers develop and implement battle plans—both offensive and defensive. They also keep in contact with other support units to coordinate plans of attack.

Special Forces personnel are trained to perform complicated and dangerous attacks from land, sea, or air, many of them at a moment's notice. They contribute to military operations by conducting offensive missions including clearing land and underwater mines and destroying key enemy targets such as bridges and railroads, ammunition stockpiles, or communication stations. They also gather information on the location and strength of enemy forces. Members of the Special Forces are highly trained in military combat as well as experts in swimming, parachuting, explosives, and survival techniques.

Special Forces officers train and supervise team members. They may oversee training needed for certain missions that involve scuba diving, intelligence gathering, or explosives detonation. When not working on an active mission, officers may lead simulated exercises, search and rescue for example, to keep their team's skills up to date.

Officers also coordinate with other Special Forces teams to plan and implement successful missions.

REQUIREMENTS

High School

Your educational preparation will depend to an extent on your career goals. At a minimum you will need a high school degree or its equivalent to join a branch of the armed forces as enlisted personnel. If you want to become an officer, however, you will also need a college education. In either case, you should take high school classes in mathematics, including advanced classes such as algebra and geometry, and science. Take computer science classes since many positions will require you to have technical skills. History classes, government classes, and classes covering geography will also be helpful. English classes will help you develop the skills to follow directions as well as to communicate clearly and precisely. Also, consider taking a foreign language, which may expand your job opportunities. Remember to take physical education classes throughout your high school years. You will be required to pass physical and medical tests when you apply to the service, so you will need to be in good physical condition.

Rank and Military Branch by Occupation

Job Title	Rank	Military Branches
Armored Assault Vehicle Crew Members	Enlisted	Army, Marines
Armored Assault Vehicle Officers	Officer	Army, Marines
Artillery and Missile Crew Members	Enlisted	Army, Marines, Navy
Artillery and Missile Officers	Officer	Air Force, Army, Coast Guard, Marines, Navy
Combat Mission Support Officers	Officer	Air Force, Army, Marines, Navy
Infantry Officers	Officer	Army, Marines
Infantry Workers	Enlisted	Army, Marines
Special Forces Officers	Officer	Air Force, Army, Marines, Navy
Special Forces Personnel	Enlisted	Air Force, Army, Marines, Navy

Source: U.S. Department of Defense

Postsecondary Training

Workers in combat specialty occupations learn how to do their jobs through both classroom and field training (including simulated combat conditions). Typical classes, depending on specialty, include instruction on ammunition handling procedures; armor offensive and defensive tactics; armor operations, principles, and tactics; artillery tactics; battle tactics and management; fire direction control procedures; gun, missile, and rocket system operations; maintenance programs; map reading; missile targeting; modern offensive and defensive combat techniques; scouting and reconnaissance techniques; security coding and authentication procedures; vehicle operations; and weapons and equipment maintenance.

Members of the infantry receive very comprehensive training. They participate in basic training for seven to eight weeks, then take advanced training in infantry skills for another eight weeks. Most of this training takes place under simulated combat conditions. Training never really stops for infantry soldiers. They must always practice their skills and stay up to date with the latest equipment, weapons, and technology.

Members of the Special Forces are taught both in the classroom and in the field. They learn a wide variety of skills such as swimming, scuba diving, parachuting, mission planning, handling and using explosives, using weapons, and offensive and defensive tactics.

Other Requirements

It takes a special individual to work in a combat specialty occupation. Those successful in their jobs are able to perform under stressful, challenging, and often dangerous situations. The ability to follow orders, react quickly, and work well as a member of a team is also important.

Officers must have leadership qualities and be able to motivate members of their units. They work well under extreme pressure, making crucial decisions that can mean life or death for hundreds, even thousands, of people.

Elite members of the Special Forces must maintain top physical health at all times. They are also skilled in such areas as scuba diving, swimming, parachuting, and survival tactics needed to complete special missions. Special Forces members often call upon their knowledge of different languages and cultures when sent to missions throughout the world.

Above all, members of the military hold their deep sense of citizenship and love of the United States as the main reason for serving their country.

EXPLORING

As a high school student, there are few ways to explore combat specialty careers firsthand. To get an idea of the physical demands of the occupations, though, you should participate in sports and stay in excellent physical shape. You can also read books about the military and visit military-oriented Web sites.

To determine a possible career path, you will need to do a fair amount of exploring to determine what job you would like to have as well as what branch of the military best suits you. Consider any family members or friends who have served in the military a valuable resource. Ask them about their experiences, what they liked best about the military life and what they liked least. Talk to recruiters from several branches to learn about what each has to offer. Attend events that are open to the public, such as air shows, where you may also have the opportunity to talk to those in the service, and visit the Web sites of each branch. (See the end of this article for contact information.) Researching before you join is one of the primary ingredients to success in this field.

EMPLOYERS

The U.S. government employs the military. In 2005, 207,574 soldiers were employed in combat specialty occupations: 6,405 individuals served in the air force; 132,524 individuals in the army; 56,918 in the marines; and 11,727 in the navy.

STARTING OUT

If you are considering entering the armed forces, contact a military recruiter. Visit the Web sites listed at the end of this article to locate a recruiting office near you. To start out in any branch, you will need to pass physical and medical tests, the Armed Services Vocational Aptitude Battery exam, and basic training.

ADVANCEMENT

Each military branch has nine enlisted grades (E-1 through E-9) and 10 officer grades (O-1 through O-10). The higher the number is, the more advanced a person's rank is. The various branches of the military have somewhat different criteria for promoting individuals; in general, however, promotions depend on factors such as length of time served, demonstrated abilities, recommendations, and scores on written exams. Promotions become more and more competitive as people advance in rank. On average, a diligent enlisted person can

Army soldiers engage Taliban forces near the village of Allah Say, Afghanistan. *(Staff Sergeant Michael L. Casteel, U.S. Army, U.S. Department of Defense)*

expect to earn one of the middle noncommissioned or petty officer rankings (E-4 through E-6); some officers can expect to reach lieutenant colonel or commander (O-5). Outstanding individuals may be able to advance beyond these levels.

EARNINGS

The U.S. Congress sets the pay scales for the military after hearing recommendations from the president. The pay for equivalent grades is the same in all services (that is, anyone with a grade of E-4, for example, will have the same basic pay whether in the army, navy, marines, air force, or Coast Guard). In addition to basic pay, personnel who frequently and regularly participate in combat may earn hazardous duty pay. Other special allowances include special duty pay and foreign duty pay. Earnings start relatively low but increase on a fairly regular basis as individuals advance in rank. See the appendix at the end of this book for detailed information on pay scales for the U.S. military. When reviewing earnings, it is important to keep in mind that members of the military receive free housing, food, and health care—items that civilians typically pay for themselves.

Additional benefits for military personnel include uniform allowances, 30 days' paid vacation time per year, and the opportunity to retire after 20 years of service. Generally, those retiring will receive 40 percent of the average of the highest three years of their base pay. This amount rises incrementally, reaching 75 percent of the average of the highest three years of base pay after 30 years of service. All retirement provisions are subject to change, however, and you should verify them as well as current salary information before you enlist. Those who retire after 20 years of service are usually in their 40s and thus have plenty of time, as well as an accumulation of skills, with which to start a second career.

WORK ENVIRONMENT

The work environment of the military changes depending on whether the United States is at war. During peacetime, those serving in combat specialty occupations are stationed at military bases in the United States, though some may be sent abroad to serve in military bases in other countries. Unit members spend this time practicing combat techniques and formations, simulated battle exercises, and different weapons systems. They must stay physically and mentally fit, and be ready for deployment at all times.

During wartime, combat specialty workers' knowledge, training, and endurance are put to the test. Infantry members, for example, establish and secure offensive camps, artillery assault vehicle members drive their tanks in combat formations, and members of the Special Forces infiltrate and weaken the enemy by destroying fuel stations or communication lines. Since combat troops fight throughout the world, members must be able to withstand extremes in climate and topography. They are often forced to sacrifice comforts such as air-conditioning, running water, or hot meals. The biggest difference in work environments, peace versus war, is the looming threat of injury, and even death.

OUTLOOK

Opportunities for combat specialty workers will be strong as a result of the continuing threat of terrorism and the presence of the U.S. military in Iraq, Afghanistan, and other countries. Those who are interested in adventure and the chance to protect their country, but also willing to put themselves at risk of serious injury or possible loss of life, will continue to have excellent opportunities in the armed forces.

FOR MORE INFORMATION

To get information on specific branches of the military, check out this Web site, which is the home of ArmyTimes.com, NavyTimes .com, AirForceTimes.com, and MarineCorpsTimes.com:

Military City
http://www.militarytimes.com

If you're thinking of joining the armed forces, take a look at this site, which guides students and parents through the decision-making process:

Today's Military
http://www.todaysmilitary.com

For information on joining the military, contact

United States Air Force
http://www.airforce.com

United States Army
http://www.goArmy.com

Unites States Coast Guard
http://www.uscg.mil

United States Marine Corps
http://www.marines.com

United States Navy
http://www.navy.com

INTERVIEW

Michael Hynes is a platoon sergeant in the U.S. Army. He discussed his career with the editors of Careers in Focus: Armed Forces.

Q. How long have you served in the military? Where have you served?

A. I have been in the military, off and on, for the last 25 years. I started out on active duty when I was a kid, moved to the Reserves for about 10 years, and (after a lengthy break in service) came into the National Guard about four years ago. I've either trained or been stationed in Germany, Saudi Arabia, Kuwait, Afghanistan, and Morocco. I've been stationed or trained in six or eight different states in the United States. Through my military service, I've been able to travel to about 20 other different countries on four continents.

Q. What made you want to enter the military?

A. Patriotism.

Q. What is one thing that young people may not know about a career in the military?

A. There is no limit on how far you can succeed in the military, as long as you are willing to play by the rules and work hard.

But to make it to the top, you need to be prepared to make a great many sacrifices in your personal life. You may spend months or years away from your family when you are deployed, and you will work long and hard hours when you are at home. A 12-hour day is standard in the U.S. Army. While on combat deployment, 20+ hour workdays are common.

Q. What are the most important personal and professional qualities for people in the military?

A. Good people skills are essential in the military, for everyone, but especially for leaders. It is shocking how many people are blinded by the idea of movies like *Rambo*. Nobody does anything in the military by themselves. The ability to work with others is key. Teamwork, teamwork, teamwork. There is no such thing as an "Army of One."

Q. What advice would you give to young people who are interested in the field?

A. There are many perks and benefits associated with military service, including college tuition, travel, adventure, and a chance to learn valuable job skills. The life of a soldier can be very, very hard, so those who enter service deserve those benefits richly. However, if the core reason for your entering military service is *not* a desire to serve the country that you love, then your military experience will be hard and difficult. If you are unable or unwilling to make sacrifices, then stay out of the Army.

Construction, Building, and Extraction Occupations

OVERVIEW

Military *construction, building, and extraction workers* build airstrips, barracks, bridges, buildings, docks, power plants, roads, water treatment plants, and other structures and facilities. Opportunities for both enlisted personnel and officers are available in the U.S. Air Force, Army, Coast Guard, Marines, and Navy. There were 32,183 soldiers employed in construction, building, and extraction occupations in 2005.

QUICK FACTS

School Subjects
Mathematics
Technical/shop

Personal Skills
Mechanical/manipulative
Technical/scientific

Work Environment
Indoors and outdoors
Primarily multiple locations

Minimum Education Level
Varies by career specialty

Salary Range
$15,617 to $30,618 to $66,154 (enlisted personnel)
$25,358 to $70,877 to $174,103 (officers)

Outlook
About as fast as the average

DOT
003, 018, 055, 840, 844, 859, 860, 861, 862, 866

GOE
02.07.04, 02.08.01, 05.03.01, 06.01.01, 06.02.01, 06.02.02, 06.02.03, 06.04.01, 08.05.01

(continues)

HISTORY

In the past, when people wanted simple structures built, they often did the work themselves. Inevitably, the larger, more complex structures required the efforts of many workers. These workers included both skilled specialists in certain activities, and others who assisted the specialists or did other less complicated but necessary physical tasks. Today, countless specialized construction careers exist—from those that require very little training (construction laborers) to those that require a bachelor's degree (civil engineers).

Wherever there has been war and a need for the military throughout history, demand has ensued for construction workers to build bridges, buildings, barracks, docks, roads, and other structures.

In the United States, construction workers and engineers have played a significant role in the military ever since the Revolutionary War. President George Washington appointed the first engineer officers of the

QUICK FACTS

(continued)

O*NET-SOC

17-2051.00, 17-3031.00, 17-3031.01, 17-3031.02, 47-2021.00, 47-2022.00, 47-2031.00, 47-2051.00, 47-2073.00, 47-2073.01, 47-2073.02, 47-2111.00, 47-2152.00, 47-2152.01, 47-2152.02, 47-2161.00, 47-2181.00

army on June 16, 1775, according to the U.S. Army Corps of Engineers, which became a permanent branch of the army in 1802.

Another famous military construction organization is the U.S. Naval Construction Force, more popularly known as the Seabees. The Naval Construction Battalions (now known as the U.S. Naval Construction Force) was founded in 1941 during World War II. The first Seabees (the name Seabees comes from the abbreviation, "C.B.," for Construction Battalions) were recruited from the civilian construction trades. Since the navy was seeking workers with experience and skill, the average age of the first Seabees was 37. More than 325,000 men served with the Seabees during World War II, according to a history of the Seabees from the U.S. Naval Construction Force. The Seabees continue to play a key role in military construction today in Iraq, Afghanistan, and other places and now also include mechanical engineers, electrical engineers, environmental engineers, and architects.

Throughout the history of the United States, construction workers and engineers in all five military branches have played a key role in war and peacetime, building bridges, for example, during the Civil War, maintaining airstrips and roads under fire in World Wars I and II, managing construction projects for U.S. allies during the Cold War, and building water treatment plants during the Iraq conflict.

THE JOB

The military offers many construction specialties. Some of the more popular career options are described below.

Building electricians design, assemble, install, test, and repair electrical fixtures and wiring. They work on a wide range of electrical and data communications systems that provide light, heat, refrigeration, air-conditioning, power, and the ability to communicate in offices, airplane hangars, repair shops, and other buildings on military bases.

Civil engineers are involved in the design and construction of the physical structures that make up buildings, roads, bridges, airfields, power plants, docks, and water treatment plants on military bases and in other locations.

Construction equipment operators handle various types of power-driven construction machines such as power shovels, cranes, derricks, hoists, pile drivers, concrete mixers, paving machines, trench excavators, bulldozers, tractors, and pumps during the construction of airfields, roads, dams, and buildings. They use these machines to move construction materials, earth, logs, coal, grain, and other materials.

Construction specialists, together with engineers and other building specialists, build many types of structures for military use, ranging from buildings, offices, bunkers, storage huts, bridges, and dams. They work on building projects from the foundation to the laying of the roofing materials. They are skilled in the use of many building materials such as lumber, concrete, masonry, and asphalt.

Plumbers and pipe fitters assemble, install, alter, and repair pipes and pipe systems that carry water, steam, air, waste, or other liquids and gases for sanitation and industrial purposes as well as other uses. Plumbers also install plumbing fixtures, appliances, and heating and refrigerating units. Pipe fitters also work on pipe systems on aircraft, missiles, and ships.

Surveying, mapping, and drafting technicians help determine, describe, and record geographic areas or features for airstrips, barracks, roads, docks, and other projects. They also create surveys and maps that are used to find military targets and plan troop movements. They operate modern surveying and mapping instruments and may participate in other operations.

Rank and Military Branch by Occupation

Job Title	Rank	Military Branches
Building Electricians	Enlisted	Air Force, Army, Coast Guard, Marines, Navy
Civil Engineers	Officer	Air Force, Army, Coast Guard, Marines, Navy
Construction Equipment	Enlisted	Air Force, Army, Coast Guard, Marines, Navy
Construction Specialists	Enlisted	Air Force, Army, Coast Guard, Marines, Navy
Plumbers and Pipe Fitters	Enlisted	Air Force, Army, Coast Guard, Marines, Navy
Surveying, Mapping, and Drafting Technicians	Enlisted	Air Force, Army, Coast Guard, Marines, Navy

Source: U.S. Department of Defense

REQUIREMENTS

High School

In high school, take courses such as mathematics (algebra, geometry, and trigonometry) and science (especially physics), shop classes that teach the use of various tools, computer science, and mechanical drawing. Electronics courses are especially important if you plan to become an electrician.

Postsecondary Training

Educational requirements in the civilian sector vary by career, but most workers receive their training by participating in an apprenticeship or by attending a technical college. Civil engineers typically train for the field by earning at least a bachelor's degree in civil engineering or a related subject.

Enlisted personnel (building electricians, construction equipment operators, construction specialists, plumbers and pipe fitters, and surveying, mapping, and drafting technicians) receive job training in classroom settings, which is sometimes augmented by on-the-job training and advanced courses. Course work varies by specialty. For example, building electricians learn how to read blueprints; use hand tools; install and wire transformers, circuit breakers, and junction boxes; and repair and replace faulty wiring and electrical fixtures. Plumbers and pipe fitters learn how to install, operate, and repair pipe systems; master soldering, silver brazing, welding, and cutting techniques; and repair and maintain pneumatic and hydraulic systems.

As officers, civil engineers must have at least a bachelor's degree in civil engineering or a related field before they join the military. Once they enter the military, they take advanced courses that will help them work on medical service and environmental control building projects.

Visit the U.S. Department of Defense's Web site, http://www.todaysmilitary.com, for more on military training in construction, building, and extraction careers.

Other Requirements

Electricians, construction specialists, and plumbers and pipe fitters should be able to use hand and power tools, work with detailed plans, and perform physical work. Construction equipment operators must be able to operate heavy equipment safely and efficiently. Because of the increasing technical nature of their work, surveying, mapping, and drafting technicians must have computer skills to be able to use highly complex equipment such as GPS and GIS technology. Civil engineers should have a passion for mathematics and science; an aptitude for problem solving, both alone and with a team; and an ability to visualize multidimensional, spatial relationships.

A U.S. Navy Seabee smoothes mortar as he builds a wall for a girls' school dormitory in Tadjoura, Djibouti. *(Chief Petty Officer Philip A. Fortnam, U.S. Navy, U.S. Department of Defense)*

Visit the U.S. Department of Defense's Web site, http://www.todaysmilitary.com, for more on personal requirements for workers in these careers.

EXPLORING

Although opportunities for direct experience in most of these occupations are rare for those in high school, there are ways to explore the field. Speaking to an experienced worker in the field will give you a clearer picture of day-to-day work in this area. Pursuing hobbies with a mechanical aspect will help you determine how much you like hands-on work. Other ways to learn about any of the jobs listed in this article include reading books and magazines about these careers, visiting Web sites of professional associations, and visiting construction sites to see workers in action.

To learn more about career opportunities in the military, visit the Web sites listed at the end of this article.

EMPLOYERS

The U.S. government employs the military. In 2005, 32,183 soldiers were employed in construction, building, and extraction occupations: 6,407 individuals served in the air force; 15,544 in the army; 5,147 in the marines; and 5,085 in the navy.

STARTING OUT

If you are thinking about joining the military, contact a military recruiter to discuss your options. Visit the Web sites listed at the end of this article to locate a recruiting office near you. To start out in any branch, you will need to pass physical and medical tests, the Armed Services Vocational Aptitude Battery exam, and basic training.

ADVANCEMENT

Each military branch has nine enlisted grades (E-1 through E-9) and 10 officer grades (O-1 through O-10). The higher the number is, the more advanced a person's rank is. The various branches of the military have somewhat different criteria for promoting individuals; in general, however, promotions depend on factors such as length of time served, demonstrated abilities, recommendations, and scores on written exams. Promotions become more and more competitive as people advance in rank. On average, a diligent enlisted person can expect to earn one of the middle noncommissioned or petty officer rankings (E-4 through E-6); some officers can expect to reach lieutenant colonel or commander (O-5). Outstanding individuals may be able to advance beyond these levels.

EARNINGS

The U.S. Congress sets the pay scales for the military after hearing recommendations from the president. The pay for equivalent grades is the same in all services (that is, anyone with a grade of E-4, for example, will have the same basic pay whether in the army, navy, marines, air force, or Coast Guard). In addition to basic pay, personnel who frequently and regularly participate in combat may earn hazardous duty pay. Other special allowances include special duty pay and foreign duty pay. Earnings start relatively low but increase on a fairly regular basis as individuals advance in rank. See the appendix at the end of this book for detailed information on pay scales for the U.S. military. When reviewing earnings, it is important to keep in mind that members of the military receive free housing, food, and health care—items that civilians typically pay for themselves.

Additional benefits for military personnel include uniform allowances, 30 days' paid vacation time per year, and the opportunity to retire after 20 years of service. Generally, those retiring will receive 40 percent of the average of the highest three years of their base pay. This amount rises incrementally, reaching 75 percent of the average of the highest three years of base pay after 30 years of service. All retirement provisions are subject to change, however, and you should verify them as well as current salary information before you enlist. Those who retire after 20 years of service are usually in their 40s and thus have plenty of time, as well as an accumulation of skills, with which to start a second career.

WORK ENVIRONMENT

Those in construction, building, and extraction occupations do demanding physical work. They may need to lift heavy weights, kneel, crouch, stoop, crawl, or work in awkward positions. They often work outdoors, sometimes in hot or cold weather, in wind or rain, in dust, mud, noise, or other uncomfortable conditions. They need to be constantly aware of danger and must be careful to observe good safety practices at all times. Often they wear gloves, hats, and vision, respiratory, or hearing protection to help avoid injury. Surveying technicians assigned to intelligence units and plumbers and pipe fitters may also work on ships.

OUTLOOK

Employment in the armed forces is expected to grow about as fast as the average for all occupations through 2014, according to the U.S. Department of Labor. When the economy is good and/or during times of war, more people pursue employment in the civilian workforce, which creates additional opportunities in the military. With the U.S. military involved in several international conflicts, most significantly in Iraq and Afghanistan, demand should continue to be strong for military workers.

Employment opportunities in this field are expected to be strong. Construction workers will always be needed in the military to help build airstrips, bridges, water treatment plants, and other structures and facilities.

Employment for workers in construction, building, and extraction occupations in the civilian sector is expected to grow about as fast as the average through 2014, according to the U.S. Department of Labor.

FOR MORE INFORMATION

For information on training programs and accreditation, contact

American Council for Construction Education
1717 North Loop 1604 East, Suite 320
San Antonio, TX 78232-1570
Tel: 210-495-6161
Email: acce@acce-hq.org
http://www.acce-hq.org

For information on training, contact

Associated General Contractors of America
2300 Wilson Boulevard, Suite 400
Arlington, VA 22201-5426
Tel: 703-548-3118
Email: info@agc.org
http://www.agc.org

For information on state apprenticeship programs, visit

Employment & Training Administration
U.S. Department of Labor
http://www.doleta.gov

For information on construction careers, contact

National Center for Construction Education and Research
3600 NW 43rd Street, Building G
Gainesville, FL 32606-8134
Tel: 888-622-3720
http://www.nccer.org

For information on engineering and construction careers in the military, contact the following organizations:

Society of Military Engineers
607 Prince Street
Alexandria, VA 22314-3117

Tel: 703-549-3800
http://www.same.org

U.S. Army Corps of Engineers
441 G Street
Washington, DC 20314-1000
http://www.usace.army.mil

U.S. Navy Civil Engineer Corps
20 Integrity Drive
Millington, TN 38055-4630
http://www.navy.com

U.S. Naval Construction Force
First Naval Construction Division
1310 8th Street, Suite 100
Norfolk, VA 23521-2435
https://www.seabee.navy.mil

To get information on specific branches of the military, check out this site, which is the home of ArmyTimes.com, NavyTimes.com, AirForceTimes.com, and MarineCorpsTimes.com:

Military City
http://www.militarytimes.com

If you're thinking of joining the armed forces, take a look at this site, which guides students and parents through the decision-making process:

Today's Military
http://www.todaysmilitary.com

For information on military careers, contact

United States Air Force
http://www.airforce.com

United States Army
http://www.goarmy.com

United States Coast Guard
http://www.gocoastguard.com

United States Marine Corps
http://www.marines.com

United States Navy
http://www.navy.com

Counseling, Social Work, and Human Services Occupations

QUICK FACTS

School Subjects
Psychology
Speech

Personal Skills
Communication/ideas
Helping/teaching

Work Environment
Indoors and outdoors
Primarily multiple locations

Minimum Education Level
Varies by career specialty

Salary Range
$15,617 to $30,618 to $66,154 (enlisted personnel)
$25,358 to $70,877 to $174,103 (officers)

Outlook
About as fast as the average

DOT
045, 120, 166, 195, 249

GOE
02.04.01, 12.02.01, 12.02.02

O*NET-SOC
13-1071.00, 13-1071.01, 19-3031.00, 19-3031.01, 19-3031.02, 19-3031.03, 19-3032.00, 21-1011.00, 21-1012.00, 21-1021.00, 21-1022.00, 21-1023.00, 21-2011.00

OVERVIEW

Military *counseling, social work, and human services workers* provide moral, emotional, and spiritual guidance and support to military personnel and their families. They help members of the military make decisions about their career paths, help families deal with the many challenges involved in living a military lifestyle, and help soldiers and their families deal with drug or alcohol abuse or emotional problems. Opportunities are available for both enlisted personnel and officers in the U.S. Air Force, Army, Coast Guard, Marines, and Navy. There were 32,610 soldiers employed in support service occupations in 2005.

HISTORY

In ancient times, people with mental health issues, substance addictions, poverty, or other personal problems were assisted by their families and friends. But as society grew more complicated and industrialized, many people began to move away from their homes to find work in other cities, and when they ran into difficulties, they were not able to count on the support of their families. Monasteries and churches filled this gap by providing charitable donations of food, clothing, and money, as well as religious support. Eventually, the government created formal programs that attempted to meet the needs of the downtrodden. Today, the social services industry—both private and governmental—is responsible for empowering individuals and helping them to face personal problems and address large social issues.

In the military, counseling and human services professionals have helped soldiers address a wide variety of issues since the early days of our country. Military chaplains, one of the most popular careers in this field, began providing religious support and counseling to troops during the Revolutionary War, and were called to assist in every subsequent war. After World War II, the position of military chaplain was formalized, and since then, religious counselors have served in the military in peacetime and during war.

THE JOB

The following paragraphs describe some of the more popular options for those interested in counseling, social work, and human services careers in the military.

Caseworkers and counselors help military personnel deal with issues ranging from substance abuse and emotional problems, to family conflicts and career challenges. They work with the individual and their families, often specializing in a particular area such as drug or alcohol abuse or social anxiety. Patients may ask for their services or be referred by their commanders. The duties of caseworkers and counselors include administering and evaluating psychological tests, interviewing and counseling patients, providing guidance regarding social or career issues, and keeping detailed records of counseling sessions. They often work with a team of social workers, psychologists, doctors, members of the clergy, and commanders to provide assistance to military personnel.

Religious counsel of all denominations is available to people in the military. *Chaplains* prepare and lead religious services for soldiers of all religious denominations, officiate at special ceremonies such as weddings and funerals, visit the sick and wounded in hospitals and on the battlefield, and offer spiritual and moral counsel to those who seek help. *Religious program specialists* are trained to help chaplains give spiritual guidance and moral support to military personnel and their families. They assist chaplains with administrative duties, religious services, and special religious meals. They are often sent on the field to give support to soldiers in war zones.

Psychologists treat the mental health needs of soldiers and their families. They also conduct research to help the military improve training, assign job duties, and design equipment. Areas of study include human and animal behavior, thinking processes, emotions, aptitude and job performance, and the relationship between equipment and

Rank and Military Branch by Occupation

Job Title	Rank	Military Branches
Caseworkers and Counselors	Enlisted	Air Force, Army, Coast Guard, Marines, Navy
Chaplains	Officer	Air Force, Army, Navy
Psychologists	Officer	Air Force, Army, Navy
Religious Program Specialists	Enlisted	Air Force, Army, Navy
Social Workers	Officer	Air Force, Army, Navy

Source: U.S. Department of Defense

technology and mental health. Psychologists also treat soldiers with mental health issues. They meet with soldiers individually or in groups to help them overcome stress, anxiety, or mental illness. They conduct psychological tests to diagnose the individual, then use a variety of treatment therapies to return them to good mental health.

Social workers help soldiers and their families solve problems. These problems include racism, discrimination, physical and mental illness, addiction, and abuse. They counsel individuals and families, lead group sessions, research social problems, and develop policy and programs.

REQUIREMENTS

High School

In high school, in addition to studying a core curriculum, with courses in English, speech, history, mathematics, and biology, you should take courses in psychology and sociology. You will also find it helpful to take business and computer science classes. If you are interested in becoming a chaplain or a religious program specialist, you should take the aforementioned classes and as many religion classes as possible.

Postsecondary Training

Educational requirements in the civilian sector vary by career. For example, most alcohol and drug abuse counselors have a bachelor's degree in counseling, psychology, health sociology, or social work, while career counselors have master's degrees. Religious workers have educational backgrounds that range from a little more than a high school education combined with short-term religious study to advanced degrees in religion. A doctorate in psychology is recommended for psychologists. Social workers have at least a bachelor's degree in social work.

Enlisted personnel in this specialty receive job training in a variety of ways, including classroom instruction, on-the-job experience, and through advanced courses. Course work varies by specialty. For example, caseworkers and counselors learn how to interview and counsel patients, perform psychological testing, and treat patients who are abusing alcohol or drugs. Religious program specialists learn basic guidance and counseling techniques, office procedures, and the principles of religious support.

Officers such as chaplains, psychologists, and social workers must enter the military with at least a bachelor's degree in their specialty. Once they join the military, they receive additional training via classroom instruction and advanced course work.

Visit the U.S. Department of Defense's Web site, http://www.todaysmilitary.com, for more on military training for those interested in counseling, social work, and human services careers.

Other Requirements

All workers in these fields must have a genuine interest in helping others, be good listeners, be sensitive to the problems of others, have patience in helping people with their problems, and be strong communicators.

Visit the U.S. Department of Defense's Web site, http://www.todaysmilitary.com, for more on personal requirements for workers in these careers.

EXPLORING

As a high school student, you may find openings for summer or part-time work as a receptionist or file clerk with a local social service agency. If no opportunity for paid employment is available, you could work as a volunteer. Good experience is also provided by working as a counselor in a camp for children with physical, mental, or developmental disabilities. Your local YMCA, park district, or other recreational facility may need volunteers for group recreation programs, including programs designed for the prevention of delinquency. You could also volunteer a few afternoons a week to read to people in retirement homes or to the blind. Work as a tutor in special education programs is sometimes available to high school students.

Others ways to learn about any of the jobs listed in this article include reading books and magazines about these careers, visiting Web sites of professional associations, arranging a tour of a social service agency, and asking your guidance counselor or teacher to arrange an information interview with a worker in the field.

To learn more about career opportunities in the military, visit the Web sites listed at the end of this article.

EMPLOYERS

The U.S. government employs the military. In 2005, 32,610 soldiers were employed in support service occupations: 2,497 individuals served in the air force; 14,963 in the army; 1,146 in the coast guard; 2,302 in the marines; and 11,702 in the navy. Approximately 3,800 military chaplains serve in the U.S. military.

STARTING OUT

If you want to enter the armed forces, contact a military recruiter. The Web sites listed at the end of this article can provide you with information on a recruiting office near you. To start out in any branch, you will need to pass physical and medical tests, the Armed Services Vocational Aptitude Battery exam, and basic training.

ADVANCEMENT

Each military branch has nine enlisted grades (E-1 through E-9) and 10 officer grades (O-1 through O-10). The higher the number is, the more advanced a person's rank is. The various branches of the military have somewhat different criteria for promoting individuals; in general, however, promotions depend on factors such as length of time served, demonstrated abilities, recommendations, and scores on written exams. Promotions become more and more competitive as people advance in rank. On average, a diligent enlisted person can expect to earn one of the middle noncommissioned or petty officer rankings (E-4 through E-6); some officers can expect to reach lieutenant colonel or commander (O-5). Outstanding individuals may be able to advance beyond these levels.

EARNINGS

The U.S. Congress sets the pay scales for the military after hearing recommendations from the president. The pay for equivalent grades is the same in all services (that is, anyone with a grade of E-4, for example, will have the same basic pay whether in the army, navy, marines, air force, or coast guard). In addition to basic pay, personnel who frequently and regularly participate in combat may earn hazardous duty pay. Other special allowances include special duty pay and foreign duty pay. Earnings start relatively low but

Two navy chaplains (far left and far right) conduct a memorial service aboard the Military Sealift Command hospital ship USNS *Comfort.* *(Mass Communication Specialist 3rd Class Tyler Jones, U.S. Navy, U.S. Department of Defense)*

increase on a fairly regular basis as individuals advance in rank. See the appendix at the end of this book for detailed information on pay scales for the U.S. military. When reviewing earnings, it is important to keep in mind that members of the military receive free housing, food, and health care—items that civilians typically pay for themselves.

Additional benefits for military personnel include uniform allowances, 30 days' paid vacation time per year, and the opportunity to retire after 20 years of service. Generally, those retiring will receive 40 percent of the average of the highest three years of their base pay. This amount rises incrementally, reaching 75 percent of the average of the highest three years of base pay after 30 years of service. All retirement provisions are subject to change, however, and you should verify them as well as current salary information before you enlist. Those who retire after 20 years of service are usually in their 40s and thus have plenty of time, as well as an accumulation of skills, with which to start a second career.

WORK ENVIRONMENT

Social workers and counselors typically work in offices or clinics. Psychologists usually work in hospitals, clinics, offices, and other medical facilities on land and aboard ships. Chaplains and religious program specialists work in hospitals, offices, and places of worship. They may also work aboard ships or aircraft or with combat units in the field. Because of this, they are sometimes exposed to dangerous and possibly life-threatening situations.

OUTLOOK

Employment in the armed forces is expected to grow about as fast as the average for all occupations through 2014, according to the U.S. Department of Labor. When the economy is good and/or during times of war, more people pursue employment in the civilian workforce, which creates additional opportunities in the military. With the U.S. military involved in several international conflicts, most significantly in Iraq and Afghanistan, demand should continue to be strong for military workers.

Demand for counseling, social work, and human services workers in the military should increase as a result of the demands that have been put on the U.S. military serving in Iraq, Afghanistan, and other countries. More soldiers will seek out these professionals to help them deal with emotional issues related to combat, injury, or the overall stress of battle and extended deployments.

The employment outlook in the civilian sector varies by occupation. The U.S. Department of Labor predicts faster than average growth for caseworkers and counselors, psychologists, and social workers; and growth about as fast as the average for chaplains.

FOR MORE INFORMATION

For more information on substance abuse and counseling careers, contact

American Counseling Association
5999 Stevenson Avenue
Alexandria, VA 22304-3300
Tel: 800-347-6647
http://www.counseling.org

For more on careers in psychology and mental health issues, contact

American Psychological Association
750 First Street, NE
Washington, DC 20002-4242
Tel: 800-374-2721
http://www.apa.org

For information on accredited training programs, contact
Association of Theological Schools in the United States and Canada
10 Summit Park Drive
Pittsburgh, PA 15275-1110
Tel: 412-788-6505
Email: ats@ats.edu
http://www.ats.edu

To access the online publication, Choices: Careers in Social Work, *contact*
National Association of Social Workers
750 First Street, NE, Suite 700
Washington, DC 20002-4241
Tel: 202-408-8600
Email: info@naswdc.org
http://www.naswdc.org

For information on a career as a military chaplain, contact
Military Chaplains Association of the United States of America
PO Box 7056
Arlington, VA 22207-7056
http://www.mca-usa.org

To get information on specific branches of the military, check out this site, which is the home of ArmyTimes.com, NavyTimes.com, AirForceTimes.com, and MarineCorpsTimes.com:
Military City
http://www.militarytimes.com

For information on military careers for officers, contact
Military Officers Association of America
http://moaa.org

If you're thinking of joining the armed forces, take a look at this site, which guides students and parents through the decision-making process:
Today's Military
http://www.todaysmilitary.com

For information on careers in the military, contact
United States Air Force
http://www.airforce.com

United States Army
http://www.goarmy.com

United States Coast Guard
http://www.gocoastguard.com

United States Marine Corps
http://www.marines.com

United States Navy
http://www.navy.com

Engineering, Science, and Technical Occupations

OVERVIEW

Military engineers are problem-solvers who use the principles of mathematics and science to plan, design, and create ways to help the military develop new weapons, equipment, and technologies. Scientific and technical workers use science and mathematics to help the military improve the performance of its troops and equipment. They study weather and ocean conditions, research ways to prevent and treat illnesses, and work to improve flight conditions for pilots, among other tasks. Opportunities are available for both enlisted personnel and officers in the U.S. Air Force, Army, Coast Guard, Marines, and Navy. There were 196,184 soldiers employed in engineering, science, and technical occupations in 2005.

HISTORY

Humankind has been "engineering," so to speak, since we realized we had opposable thumbs that we could use to handle tools. And from that point on we began our ceaseless quest to make, build, and create tools and systems that helped us live our lives better. There were a lot of mistakes, but we learned from them and built a foundation of engineering laws and principles. Today, dozens of engineering specialties exist—from aerospace engineering to software engineering.

QUICK FACTS

School Subjects
Computer science
Mathematics
Physics

Personal Skills
Leadership/management
Technical/scientific

Work Environment
Indoors and outdoors
Primarily multiple locations

Minimum Education Level
Varies by career specialty

Salary Range
$15,617 to $30,618 to $66,154 (enlisted personnel)
$25,358 to $70,877 to $174,103 (officers)

Outlook
About as fast as the average

DOT
002, 012, 015, 022, 023, 024, 025, 041, 099

GOE
02.01.01, 02.02.01, 02.03.01, 02.03.02, 02.03.03, 02.03.04, 02.05.01, 02.07.01, 02.07.03, 02.07.04

O*NET-SOC
17-2011.00, 17-2071.00, 17-2072.00, 17-2112.00,

(continues)

QUICK FACTS

(continued)

17-2121.00, 17-2161.00, 19-1020.01, 19-1021.00, 19-1021.02, 19-2012.00, 19-2021.00, 19-2031.00, 19-2042.01

Throughout human history, discoveries in the life and physical sciences have profoundly affected daily life and the overall quality of life. From the establishment of biology as a basic science by Aristotle in ancient Greece, to the invention of the thermometer by Galileo in 1593, to the first self-sustained nuclear chain reaction in 1942 by the physicist Enrico Fermi, to developments in biotechnology today, life and physical scientists have played an integral role in the development of our civilization.

Space exploration only began in the second half of the 20th century, but its roots can be traced back to 1926 with the launch of the first space vehicle propelled by ejection of gases, the rocket. In America, the National Advisory Committee for Aeronautics (NACA), founded in 1915, pursued research in rocketry with an eye toward developing rockets for use as weapons as well as to put humans into space. In 1958, the United States reorganized and expanded its space exploration efforts. The National Aeronautics and Space Administration (NASA) began operations in October 1958. It absorbed NACA and all of its resources as well as already existing military operations with space laboratories, including the Naval Research Laboratory in Maryland, the Jet Propulsion Laboratory managed by the California Institute of Technology for the army, and the Army Ballistic Missile Agency in Huntsville, Alabama. NASA began conducting space missions shortly after its creation, answering to early Soviet success with the launch of the first man-made Earth satellite, *Sputnik*. In 1961, astronaut Alan Shepard became the first American to travel in space. A series of other firsts in space followed in the 1960s: John Glenn became the first American to orbit Earth in February 1962; in 1964, the Soviets placed in orbit the first spacecraft that carried more than one person; Russian cosmonaut Alexei Leonov became the first person to step outside a spacecraft in 1965; American astronauts Neil Armstrong and Edwin Aldrin Jr. became the first people to set foot on the moon in 1969. Today, developments, such as the reusable space shuttle, the Hubble Space Telescope, the International Space Station (a permanent orbiting laboratory in space), the Mars Exploration Program, and the new Crew Exploration Vehicle (which will replace the shuttle by 2011) have renewed our ambitions toward exploring space.

Rank and Military Branch by Occupation

Job Title	Rank	Military Branches
Aerospace Engineers	Officer	Air Force, Army, Coast Guard, Marines, Navy
Electrical and Electronics Engineers	Officer	Air Force, Army, Coast Guard, Marines, Navy
Industrial Engineers	Officer	Air Force, Army, Coast Guard, Marines, Navy
Life Scientists	Officer	Air Force, Army, Navy
Marine Engineers	Officer	Coast Guard
Meteorological Specialists	Enlisted	Air Force, Army, Coast Guard, Marines, Navy
Nuclear Engineers	Officer	Army, Marines, Navy
Physical Scientists	Officer	Air Force, Army, Coast Guard, Marines, Navy
Space Operations Officers	Officer	Air Force, Army, Marines, Navy
Space Operations Specialists	Enlisted	Air Force, Navy
Unmanned Vehicle Operations Specialists	Enlisted	Air Force, Army, Marines, Navy

Source: U.S. Department of Defense

Engineers and scientific research workers have played a significant role in the military ever since the Revolutionary War. President George Washington appointed the first engineer officers of the army on June 16, 1775, according to the U.S. Army Corps of Engineers, which was made a permanent branch of the army in 1802.

Engineers and scientists played a crucial role in World War II, constructing buildings and bridges, designing new weaponry, developing equipment that was easier to use and more comfortable, tracking weather conditions, and, most significantly, developing the atomic bomb, which would finally end the war.

Throughout modern history, military engineers and scientists have been played a crucial role in peace and in wartime. Today, military engineers are responsible for developing more effective weapons and protective armor, studying diseases in underdeveloped countries, designing new Stealth aircraft, working to find ways to reduce the harmful effects of chemical or radiological attacks on soldiers, helping in the rebuilding efforts in Iraq and Afghanistan, and, through work in the U.S. Army Corps of Engineers, protecting water resources for both civilian and military populations.

THE JOB

Many engineering and scientific research specialties exist in the military. The following paragraphs describe some of the more popular career options.

Aerospace engineering encompasses the fields of aeronautical (aircraft) and astronautical (spacecraft) engineering. Military aerospace equipment, such as fighter jets, missiles, and spacecraft, is built by private contractors, but *aerospace engineers* in the military still play an integral role in their production. They conduct research on aircraft propulsion, guidance, and weapons systems; help select private contractors to build this equipment; monitor the production of aircraft, missiles, and spacecraft; determine testing processes for prototypes; and conduct wind tunnel and stress analysis tests on aircraft and missile prototypes. Some aerospace engineers specialize in overseeing the design and manufacture of one complete machine, perhaps a new jet fighter, whereas others focus on separate components such as for missile guidance systems.

Electrical and electronics engineers apply their knowledge of the sciences and engineering to create electrical and electronic systems for radar, missile guidance systems, communication equipment, and other systems. In addition to designing these systems, they are responsible for their testing, installation, and repair.

Industrial engineers use their knowledge of various disciplines—including systems engineering, management science, operations research, and fields such as ergonomics—to determine the most efficient and cost-effective methods for industrial production of high-quality military products and systems. They are responsible for designing systems that integrate materials, equipment, information, and people in the overall production process.

Life scientists study the cause and effects of different diseases on humans and animals to find treatment and prevention. Their duties include studying the effects of different drugs, poisons, chemicals, bacteria, or poisons on soldiers in hopes of offsetting any negative impacts. They also study different food storage and handling methods to avoid contamination or illness. Life scientists may also find ways to better keep military bases free from pests or contagious diseases. Some scientists may direct the operation of blood banks, including the study of blood chemistry.

Marine engineers make design modifications to different watercrafts to help them increase in speed, strength, durability, and safety. They study the design and build of a submarine's hull, a ship's deck, or equipment; make recommendations; and oversee any repairs or construction. Marine engineers also work with the ship's combat and salvage equipment and make changes to improve its operation.

Meteorological specialists study weather conditions and forecast weather changes to help the military plan its operations, including troop movements, ship traffic, and airplane flights. By analyzing weather maps covering large geographic areas and related charts, like upper-air maps and soundings, meteorological specialists can predict the movement of fronts, precipitation, and pressure areas. They forecast such data as temperature, winds, precipitation, cloud cover, and flying conditions. To predict future weather patterns and to develop increased accuracy in weather study and forecasting, meteorologists conduct research on such subjects as atmospheric electricity, clouds, precipitation, hurricanes, and data collected from weather satellites. Other areas of research used to forecast weather may include ocean currents and temperature.

Nuclear engineers are concerned with accessing, using, and controlling the energy released when the nucleus of an atom is split. The process of splitting atoms, called fission, produces a nuclear reaction, which creates radiation in addition to nuclear energy. Nuclear engineers in the military design, develop, and operate nuclear power plants, which are used to generate electricity and power navy ships. Others specialize in developing nuclear technology that is used in weapons and defense systems.

Physical scientists conduct research and experiments in many fields to support military missions as well as develop new weapons, methods of weather forecasting, or biological weapons. Their research might determine better ways of guarding against the effects of possible chemical or nuclear warfare or ways to streamline the design of a submarine-launched missile system. Physical scientists work in many specialties including chemistry, physics, meteorology, and oceanography. Other duties include preparing reports about their research for department heads, training other military personnel, and managing the work of laboratory workers or field staff.

The military often uses satellites to help with military missions such as enemy surveillance or base-to-base communications. *Space operations specialists* operate and maintain ground control command equipment, such as mobile transporters, test chambers, and simulators, used by spacecrafts. Their duties include transmitting spacecraft commands using aerospace ground equipment, monitoring and processing data gathered from computers and telemetry display systems, and tracking spacecrafts. They also perform routine inspections and repair ground and aerospace communication systems.

Space operations officers supervise the work done for space flight planning and training, including the launch and recovery of spacecrafts. Their duties include managing of guidance, navigation, and propulsion systems of ground and space vehicles. They work closely with astronauts and mission control facility members in planning space flights. Simulation exercises, operational tasks, and aerospace experiments are planned and implemented by space operations officers. They also monitor foreign space activity such as space flights and missile launches.

The military often uses self-piloted or control-navigated vehicles for missions that are extremely dangerous for human operators, or when the terrain is too remote. *Unmanned vehicle operations specialists* are trained in the operation, maintenance, and repair of such vehicles. They often specialize in the type of vehicle—aerial, ground, or underwater. For example, air force specialists self-pilot unmanned aerial vehicles, which carry cameras and satellites for surveillance. Army specialists control unmanned ground vehicles such as armed robotic vehicles designed to deploy sensors and munitions into buildings and tunnels.

REQUIREMENTS

High School

While in high school, follow a college preparatory program. Doing well in mathematics and science classes is vital if you want to pursue a career in any type of engineering or science field. The American Society for Engineering Education advises students to take calculus and trigonometry in high school, as well as laboratory science classes. Such courses provide the skills you'll need for problem solving, which is essential in any type of engineering. If you are interested in pursuing careers in the life, physical, or meteorological sciences, you should take as many sciences classes as possible, including biochemistry, biology, microbiology, and earth science.

Postsecondary Training

Educational requirements in the civilian sector vary by career, but most workers have at least a bachelor's degree in engineering or a science-related major.

In the military, enlisted personnel (meteorological specialists, space operations specialists, and unmanned vehicle operations specialists) train via classroom instruction and advanced course work.

Course work varies by specialty. For example, meteorological specialists learn the basics of meteorology and oceanography, techniques to plot weather data, and how to analyze radar and satellite weather data in order to prepare weather forecasts.

Officers (engineers, life and physical scientists, and space operations officers) must enter the military with at least a bachelor's degree in their field. Once they join the military, some workers (such as electrical and electronics engineers and industrial engineers) receive additional specialized training via classroom instruction, on-the-job experience, and advanced course work. For example, electrical and electronics engineers receive additional training in combat and tactical systems and networks and weapon system electronics. Space operations officers participate in one year of classroom instruction and practical experience. They continue their training on the job and by taking classes.

Visit the U.S. Department of Defense's Web site, http://www.todaysmilitary.com, for more on military training in engineering, science, and technical careers.

Other Requirements

Personal requirements for workers in these occupations vary by specialty. Engineers should enjoy solving problems, developing logical plans, and designing things. They should have a strong interest and ability in science and mathematics. Engineers often work on projects in teams, so prospective engineers should be able to work well both alone and with others. Life and physical scientists must have good communication skills, be organized, enjoy conducting research, and have comprehensive knowledge about their field. Space operations officers and specialists should be very interested in space exploration, be decisive, have good organizational skills, and be able to work as a member of a team. Unmanned vehicle operations specialists should be proficient with tools, be able to work as a member of a team, be knowledgeable about electronic theory and schematic drawing, and understand three-dimensional spatial relationships.

Visit the U.S. Department of Defense's Web site, http://www.todaysmilitary.com, for more on personal requirements for workers in these careers.

EXPLORING

Perhaps the best way for high school students to explore the field of engineering is by contacting the Junior Engineering Technical Society (JETS). JETS can help students learn about different fields within engineering and can guide them toward science and engineering fairs. Participation in science and engineering fairs can be an invaluable experience for a high school student interested in these fields.

Others ways to learn about any of the jobs listed in this article include reading books and magazines about engineering, science, and space exploration; visiting Web sites of professional associations; arranging a tour of an engineering firm, laboratory, or National Aeronautics and Space Administration facilities; and asking your guidance counselor or teacher to arrange an information interview with a worker in the field (such as a nuclear engineer or life scientist).

To learn more about career opportunities in the military, visit the Web sites listed at the end of this article.

EMPLOYERS

The U.S. government employs the military. In 2005, 196,184 soldiers were employed in engineering and scientific research occupations: 68,235 individuals served in the air force; 54,979 in the army; 2,079 in the Coast Guard; 19,162 in the marines; and 51,729 in the navy.

STARTING OUT

Contact a military recruiter for more information on careers in the armed forces. Visit the Web sites listed at the end of this article to locate a recruiting office near you. To start out in any branch, you will need to pass physical and medical tests, the Armed Services Vocational Aptitude Battery exam, and basic training.

ADVANCEMENT

Each military branch has nine enlisted grades (E-1 through E-9) and 10 officer grades (O-1 through O-10). The higher the number is, the more advanced a person's rank is. The various branches of the military have somewhat different criteria for promoting individuals; in general, however, promotions depend on factors such as length of time served, demonstrated abilities, recommendations, and scores on written exams. Promotions become more and more competitive as people advance in rank. On average, a diligent enlisted person can expect to earn one of the middle noncommissioned or petty officer rankings (E-4 through E-6); some officers can expect to reach lieutenant colonel or commander (O-5). Outstanding individuals may be able to advance beyond these levels.

American Troops Are Everywhere

As of June 30, 2007, 1,082,627 active duty military personnel were stationed in the United States. In addition to serving in the United States, military workers are deployed throughout the world. In fact, they are stationed in more than 150 countries. Here are the foreign countries that have the most U.S. troops.

1. Iraq: 202,100 troops
2. Germany: 58,894 troops
3. Japan: 33,068 troops
4. Republic of Korea: 27,114 troops
5. Afghanistan: 24,800 troops
6. Italy: 10,216 troops
7. United Kingdom: 10,152 troops
8. Djibouti: 2,038 troops
9. Turkey: 1,668 troops
10. Serbia (including Kosovo): 1,395 troops

Source: U.S. Department of Defense

EARNINGS

The U.S. Congress sets the pay scales for the military after hearing recommendations from the president. The pay for equivalent grades is the same in all services (that is, anyone with a grade of E-4, for example, will have the same basic pay whether in the army, navy, marines, air force, or Coast Guard). In addition to basic pay, personnel who frequently and regularly participate in combat may earn hazardous duty pay. Other special allowances include special duty pay and foreign duty pay. Earnings start relatively low but increase on a fairly regular basis as individuals advance in rank. See the appendix at the end of this book for detailed information on pay scales for the U.S. military. When reviewing earnings, it is important to keep in mind that members of the military receive free housing, food, and health care—items that civilians typically pay for themselves.

Additional benefits for military personnel include uniform allowances, 30 days' paid vacation time per year, and the opportunity to retire after 20 years of service. Generally, those retiring will receive 40 percent of the average of the highest three years of their base pay.

This amount rises incrementally, reaching 75 percent of the average of the highest three years of base pay after 30 years of service. All retirement provisions are subject to change, however, and you should verify them as well as current salary information before you enlist. Those who retire after 20 years of service are usually in their 40s and thus have plenty of time, as well as an accumulation of skills, with which to start a second career.

WORK ENVIRONMENT

Engineers usually have a central office or laboratory from which they base their work, and these offices are typically pleasant, clean, and climate controlled. Most engineers, however, are required to spend at least part of their time on a specific worksite, and these sites may be noisy, dusty, dirty, and unpleasant. Engineers may find themselves at construction sites and "hard hat" areas, on assembly lines and runways, at sewage treatment plants, at shipyards, in underground culverts and pipelines, and at machine shops and drill rigs. In addition, to working in offices, power plant control centers, and laboratories, nuclear engineers work on nuclear-powered ships and submarines.

Nuclear engineers will encounter two unique concerns. First, exposure to high levels of radiation may be hazardous; thus, engineers must always follow safety measures. Those working near radioactive materials must adhere to strict precautions outlined by regulatory standards. In addition, female engineers of childbearing age may not be allowed to work in certain areas or perform certain duties because of the potential harm to the human fetus from radiation.

Meteorological specialists typically work in offices on land or aboard ships. They work outside when making weather observations, launching weather balloons, and conducting other research.

Life scientists work in clinical, medical, and research laboratories. Others work in food-processing or storage plants. They may work outdoors in sometimes demanding conditions, including on ships, while conducting field research.

Physical scientists work in laboratories and offices on ships and on land. Others, depending on their specialization, spend many hours conducting research in the field.

Some life and physical scientists work with toxic substances and disease cultures; strict safety measures must be observed.

Space operations professionals work in offices and space operations centers. Unmanned vehicle operations specialists have a variety of work settings depending on the type of vehicle and mission. These include control stations or receiving stations on land and on ship.

OUTLOOK

Employment in the armed forces is expected to grow about as fast as the average for all occupations through 2014, according to the U.S. Department of Labor. When the economy is good and/or during times of war, more people pursue employment in the civilian workforce, which creates additional opportunities in the military. With the U.S. military involved in several international conflicts, most significantly in Iraq and Afghanistan, demand should continue to be strong for military workers.

Opportunities should be strong for workers in this field. The military would not be able to retain its technological edge over foreign armies without the work of its engineers and scientists.

The employment outlook in the civilian sector varies by occupation. The U.S. Department of Labor provides the following predictions for professions in the field: aerospace engineers, life scientists, marine engineers, and nuclear engineers—more slowly than the average; electrical and electronics engineers, industrial engineers, meteorological specialists, physical scientists, and space operations professionals—about as fast as the average.

FOR MORE INFORMATION

For a list of accredited engineering schools and colleges, contact

Accreditation Board for Engineering and Technology
111 Market Place, Suite 1050
Baltimore, MD 21202-7116
Tel: 410-347-7700
http://www.abet.org

For general information about chemistry careers and approved education programs, contact

American Chemical Society
1155 16th Street, NW
Washington, DC 20036-4801
Tel: 800-227-5558
http://www.chemistry.org

For information on careers in biology, contact

American Institute of Biological Sciences
1444 I Street, NW, Suite 200
Washington, DC 20005-6535
Tel: 202-628-1500
http://www.aibs.org

For employment statistics and information on physics jobs and career planning, contact

American Institute of Physics
One Physics Ellipse
College Park, MD 20740-3843
Tel: 301-209-3100
Email: aipinfo@aip.org
http://www.aip.org

For information on careers, education, and scholarships, contact

American Meteorological Society
45 Beacon Street
Boston, MA 02108-3693
Tel: 617-227-2425
Email: amsinfo@ametsoc.org
http://www.ametsoc.org

For information on educational programs and to purchase a copy of Engineering: Go For It, *contact*

American Society for Engineering Education
1818 N Street, NW, Suite 600
Washington, DC 20036-2479
Tel: 202-331-3500
http://www.asee.org

Visit the ASLO's Web site for information on careers and education.

American Society of Limnology and Oceanography (ASLO)
5400 Bosque Boulevard, Suite 680
Waco, TX 76710-4446
Tel: 800-929-2756
Email: business@aslo.org
http://www.aslo.org

For information on careers and student competitions, contact

Junior Engineering Technical Society
1420 King Street, Suite 405
Alexandria, VA 22314-2794
Tel: 703-548-5387
Email: info@jets.org
http://www.jets.org

For information on aeronautical careers, internships, and student projects, contact the information center or visit NASA's Web site.

National Aeronautics and Space Administration (NASA)
Public Communications and Inquiries Management Office
Washington, DC 20546-0001
Tel: 202-358-0001
Email: public-inquiries@hq.nasa.gov
http://www.nasa.gov

For information on engineering careers in the military, contact
Society of Military Engineers
607 Prince Street
Alexandria, VA 22314-3117
Tel: 703-549-3800
http://www.same.org

U.S. Army Corps of Engineers
441 G Street
Washington, DC 20314-1000
http://www.usace.army.mil

To get information on specific branches of the military, check out this site, which is the home of ArmyTimes.com, NavyTimes.com, AirForceTimes.com, and MarineCorpsTimes.com:
Military City
http://www.militarytimes.com

If you're thinking of joining the armed forces, take a look at this site, which guides students and parents through the decision-making process:
Today's Military
http://www.todaysmilitary.com

For information on military careers, contact
United States Air Force
http://www.airforce.com

United States Army
http://www.goarmy.com

United States Coast Guard
http://www.gocoastguard.com

United States Marine Corps
http://www.marines.com

United States Navy
http://www.navy.com

Health Care Occupations

QUICK FACTS

School Subjects
Biology
Chemistry
Health

Personal Skills
Helping/teaching
Technical/scientific

Work Environment
Indoors and outdoors
Primarily multiple locations

Minimum Education Level
Varies by career specialty

Salary Range
$15,617 to $30,618 to $66,154 (enlisted personnel)
$25,358 to $70,877 to $174,103 (officers)

Outlook
About as fast as the average

DOT
070, 072, 075, 076, 077, 078, 079, 187, 354

GOE
02.03.01, 04.01.01, 04.04.01, 08.02.02, 09.07.02, 14.01.01, 14.02.01, 14.03.01, 14.04.01, 14.05.01, 14.06.01, 14.07.01, 14.08.01

(continues)

OVERVIEW

The military relies on *health care workers* to care for the health and physical well-being of soldiers and their families. Opportunities are available for both enlisted personnel and officers in the U.S. Air Force, Army, Coast Guard, and Navy. There were 98,482 soldiers employed in health care occupations in 2005.

HISTORY

The origins of medicine began with prehistoric people who believed that diseases were derived from supernatural powers. To destroy the evil spirits, they performed trephining, which involved cutting a hole in the victim's skull to release the spirit. Skulls have been found in which the trephine hole has healed, demonstrating that people did survive the ritual. The first doctors, known as medicine men, also used herbal concoctions, ritual dances, and incantations to heal their patients.

Society has come a long way from these early attempts to treat the sick and injured. Some of the most groundbreaking developments in health care in recent centuries include development of the first vaccinations that battled or prevented diseases in the mid-1800s; invention of anesthesia in 1846; development of drugs such as penicillin in 1928 that battled and killed some bacteria and infections; development of the first vaccine for influenza in 1945; first use of a heart-lung machine in 1953; first human heart transplant in 1967; first implantation of an artificial heart in 1982; and mapping of the Human Genome in 2003.

Today, the health care industry continues to develop at a rapid rate with the discoveries of new drugs, treatments, and cures. Modern technologies, such as computers and virtual reality, are being used by the medical community to perform tests, compile data, diagnose illnesses, and train professionals. Many surgeries are no longer performed with a scalpel, but with lasers. Disease, illness, and injury are now being treated and cured so successfully that the general population is living much longer and the number of elderly people is increasing.

QUICK FACTS

(continued)

O*NET-SOC
11-9111.00, 29-1021.00, 29-1031.00, 29-1041.00, 29-1061.00, 29-1062.00, 29-1063.00, 29-1064.00, 29-1065.00, 29-1066.00, 29-1067.00, 29-1069.99, 29-1071.00, 29-1111.00, 29-1121.00, 29-1122.00, 29-1123.00, 29-1127.00, 29-2021.00, 29-2071.00, 31-1012.00, 31-2011.00, 31-2012.00, 31-2021.00, 31-9091.00, 51-9081.00

Health care workers have provided medical care to soldiers for as long as wars have been fought and blood has been shed. According to Militarytimes.com, the formal history of health care in the U.S. military dates back more than 200 years. Congress enacted legislation that called for care of the "regimental sick" and care for the "relief of sick and disabled seamen." These services were eventually extended to the families of soldiers as well as retired military personnel.

Today, U.S. military health care is considered the best in the world. Advances in technology and the professionalism of its workers keep military personnel healthy during times of peace and war.

THE JOB

Many health care specialties exist in the military. The following paragraphs describe some of the more popular career options.

Dental specialists perform a variety of duties in military clinics throughout the world, including helping the dentist examine and treat patients and completing laboratory and office work. They assist the dentist by preparing patients for dental exams, handing the dentist the proper instruments, taking and processing X rays, preparing materials for making impressions and restorations, and instructing patients in oral health care. They also perform administrative and clerical tasks so that the office runs smoothly and the dentist's time is available for working with patients. Dental specialists also perform oral prophylaxis, a process of cleaning teeth by using sharp dental instruments, such as scalers and prophy angles. With these instruments, they remove stains and calcium deposits, polish teeth, and massage gums.

Dentists maintain military workers' teeth through such preventive and reparative practices as extracting, filling, cleaning, and replacing teeth. They perform corrective work, such as straightening teeth, and treat diseased gum tissue. They also perform surgical operations on the jaw or mouth, and make and fit false teeth.

Dental and optical laboratory technicians make and repair devices for military personnel as prescribed by optometrists and dentists. For example, based on the measurements provided by the dentist, dental technicians construct dentures, including hardening and curing the material in powerful ovens. Finally, they grind and polish the dentures until smooth. Optical technicians construct eyewear lenses, for example, using heat-treating equipment and finishing the lenses with small hand tools before slipping them into a frame.

Dietitians provide information on nutrition to patients and outpatients in military hospitals. They establish policies for hospital food service, create special diets for patients when requested by physicians and other health care professionals, plan hospital menus, develop budgets, inspect food preparation areas to ensure they meet sanitation and safety standards, and provide information on nutrition to the military.

Medical care technicians help doctors, nurses, and other health care workers care for sick or wounded soldiers. They take patients' vital signs; feed, wash, and dress patients; take patients to other areas of the hospital for treatment, therapy, or diagnostic testing; give medicines to patients under the supervision of physicians and nurses; and prepare patients, equipment, operating rooms, and supplies for surgery and other medical procedures.

Medical service technicians treat injured or wounded soldiers during emergencies or during combat when physicians are unavailable. They often accompany the patient to a field hospital or triage area, where the individual can receive more advanced treatment from a physician or surgeon. Medical service technicians perform emergency first aid, take vital signs, interview patients about their medical histories, and keep health records and files up to date.

Cardiopulmonary and electroencephalograph (EEG) technicians help doctors diagnose and treat disorders of the heart, lung, and brain by administering noninvasive tests that look for elevations or abnormalities including blood oxygen levels and compromised functions. Their duties include operating equipment for electrocardiograms, electroencephalograms, or other tests. They take patients' blood pressure, adjust settings, and monitor the equipment throughout the duration of the test. They keep records of the test, and discuss results with attending physicians.

Optometric technicians help optometrists provide vision care to members of the military. They assist optometrists during vision exams by performing screening tests, applying eye dilation drops, and adjusting or repairing eyeglasses. They are trained in the proper use of ophthalmic instruments used during eye exams such as corneal topographers or auto-refractors. They also place orders for eyeglasses or contact lenses and take care of patients' charts.

Optometrists provide primary eye care services. These include conducting comprehensive eye health and vision examinations; diagnosing and treating eye diseases and vision disorders; prescribing glasses, contact lenses, vision therapy, and medications; performing minor surgical procedures; and counseling patients regarding their vision needs. While examining a patient's eyes, optometrists may also identify signs of diseases and conditions that affect the entire body.

Physical therapists are health care specialists who restore mobility, alleviate pain and suffering, and work to prevent permanent disability for military personnel who have been disabled by injury or illness. They test and measure the functions of the musculoskeletal, neurological, pulmonary, and cardiovascular systems. Physical therapists provide preventive, restorative, and rehabilitative treatment for their patients.

Occupational therapists use a wide variety of activities to help patients who have been injured attain their goals for productive, independent living. These goals include developing maximum self-sufficiency in activities of daily living, such as eating, dressing, writing, using a telephone and other communication resources, as well as functioning in the community and the workplace.

Physical therapy specialists help to restore physical function in those with injury or disease. They assist physical therapists with a variety of techniques, such as exercise, massage, heat, and water therapy. They teach and help patients improve functional activities required in their daily lives, such as walking, climbing, and moving from one place to another. Specialists observe patients during treatments, record the patients' responses and progress, and report these to the physical therapist, either orally or in writing. They fit patients for and teach them to use braces, artificial limbs, crutches, canes, walkers, wheelchairs, and other devices. They may make physical measurements to assess the effects of treatments or to evaluate patients' range of motion, length and girth of body parts, and vital signs.

Occupational therapy specialists help people with mental, physical, developmental, or emotional limitations using a variety of activities to improve basic motor functions and reasoning abilities. They work under the direct supervision of an occupational therapist, and their duties include helping to plan, implement, and evaluate rehabilitation programs designed to regain patients' self-sufficiency and to restore their physical and mental functions.

Physician assistants help physicians provide medical care to patients. They may be assigned a variety of tasks including taking medical histories of patients, conducting routine physical examinations, ordering laboratory tests, drawing blood samples, giving injections, diagnosing illnesses, choosing treatments, and assisting in surgery.

Physicians diagnose, prescribe medicines for, and otherwise treat diseases and disorders of the human body. A physician may also perform surgery and often specializes in one aspect of medical care and treatment. *Surgeons* are physicians who make diagnoses and provide preoperative, operative, and postoperative care in surgery affecting almost any part of the body. These doctors also work with trauma victims, such as soldiers injured by landmines, bullet wounds, improvised explosive devices, and booby traps.

Registered nurses help care for the sick and injured in hospitals, clinics, and other military facilities. They take patients' vital signs, administer medication and injections, record the symptoms and progress of patients, change dressings, assist patients with personal care, confer with members of the medical staff, help prepare patients for surgery, and complete any number of duties that require skill and understanding of patients' needs.

Speech therapists help people who have speech and hearing problems. They identify the problem, then use tests to further evaluate it. Speech therapists try to improve the speech and language skills of patients with communications disorders. These problems may have been caused by traumatic brain injury, gunshot wounds, and other injuries inflicted during battle.

Health services administrators plan and direct the delivery of health care at military hospitals, clinics, and health care facilities. They manage the day-to-day operations of key medical departments such as nursing, as well as support departments including food services and maintenance. Administrators develop and implement a working budget for health care facilities and their programs. They often meet with the heads of all hospital departments to discuss possible improvements of in-patient and out-patient care. Their duties include hiring and evaluating personnel, conducting in-service programs, and record-keeping.

Medical record technicians organize, file, and maintain records for all patients receiving medical care at military hospitals, clinics, and other health care facilities. They complete patients' admission and discharge papers, prepare and file reports for examinations, illnesses,

Rank and Military Branch by Occupation

Job Title	Rank	Military Branches
Cardiopulmonary and EEG Technicians	Enlisted	Air Force, Army, Coast Guard, Navy
Dental and Optical Laboratory Technicians	Enlisted	Air Force, Army, Coast Guard, Navy
Dental Specialists	Enlisted	Air Force, Army, Marines, Navy
Dentists	Officer	Air Force, Army, Coast Guard, Navy
Dietitians	Officer	Air Force, Army, Navy
Health Services Administrators	Officer	Air Force, Army, Coast Guard, Marines, Navy
Medical Care Technicians	Enlisted	Air Force, Army, Coast Guard, Navy
Medical Laboratory Technicians	Enlisted	Air Force, Army, Coast Guard, Marines, Navy
Medical Record Technicians	Enlisted	Air Force, Army, Coast Guard, Navy
Medical Service Technicians	Enlisted	Air Force, Army, Coast Guard, Navy
Optometric Technicians	Enlisted	Air Force, Navy
Optometrists	Officer	Air Force, Army, Navy
Physical and Occupational Therapists	Officer	Air Force, Army, Coast Guard, Navy
Physical and Occupational Therapy Specialists	Enlisted	Air Force, Army, Coast Guard, Navy
Physician Assistants	Officer	Air Force, Army, Coast Guard, Navy
Physicians and Surgeons	Officer	Air Force, Army, Coast Guard, Navy
Registered Nurses	Officer	Air Force, Army, Coast Guard, Navy
Speech Therapists	Officer	Air Force, Army, Coast Guard, Navy

Source: U.S. Department of Defense

and treatments. Their duties include making room assignments for patients and maintaining publications in the medical library.

REQUIREMENTS

High School

If you are interested in a career in health care, take college preparatory classes while you are in high school. Science courses, especially those involving laboratory work, such as biology and chemistry, will be particularly helpful. Be sure to take math classes, including algebra and calculus, and computer science courses. Round out your education with humanities classes, including English. English courses will give you the opportunity to develop your research and report writing skills, as well as your communication skills.

Postsecondary Training

Educational requirements in the civilian sector vary by career. For example, optometrists must have a doctorate degree, while physicians and surgeons must have a medical degree. Physical therapists need a master's degree. An associate's degree is the minimum educational requirement for dental hygienists. Contact the professional associations listed at the end of this article for more information on education requirements for civilian workers.

In the military, enlisted personnel (dental specialists, dental and optical laboratory technicians, medical care technicians, medical service technicians, cardiopulmonary and electroencephalograph technicians, optometric technicians, physical and occupational therapy specialists, and medical record technicians) learn how to do their jobs via classroom instruction, on-the-job training, and advanced course work. Course work and other training varies by specialty. For example, dental specialists learn about preventive dentistry, how to take X rays, and the basics of dental hygiene. They also gain experience by working with actual patients. Medical service technicians are trained in emergency medical treatment, basic nursing care, and clinical laboratory procedures. Physical therapists learn about anatomy, physiology, and psychology; therapy methods; and the handling and positioning of patients.

Officers (dentists, dietitians, optometrists, physical and occupational therapists, physician assistants, physicians, surgeons, registered nurses, speech therapists, and health services administrators) must enter the military with at least a bachelor's degree in their field. Physicians, surgeons, and dentists must have a medical degree; optometrists must have a doctorate; physical therapists and speech therapists must have a master's degree. Once they join the military, some workers (such as physician assistants, physicians and surgeons, and registered nurses) receive additional specialized training via classroom instruction and advanced course work.

Visit the U.S. Department of Defense's Web site, http://www.todaysmilitary.com, for more on military training for those interested in health care careers.

Other Requirements

Personal requirements for workers in these occupations vary by specialty. For example, dental hygienists must have skill in handling delicate instruments, a sensitive touch, and good depth perception. Dietitians must be detail oriented, able to think analytically, and be comfortable making decisions and acting on them. Physician assistants should be able to work well with many different kinds of people, from the physician who supervises them to the many different patients they see every day. They must also have a strong desire to continue learning to keep up with the latest medical procedures. Patience, compassion, decisiveness, and a desire to help others are qualities needed by anyone working in these careers.

Visit the U.S. Department of Defense's Web site, http://www.todaysmilitary.com, for more on personal requirements for workers in these careers.

EXPLORING

You can explore your interest in health care in a number of ways. Read books on careers in health care. Visit hospitals and clinics to observe the work and talk with hospital personnel to learn more about the daily activities of health care staff.

Some hospitals now have extensive volunteer service programs in which high school students may work after school, on weekends, or during vacations in order to both render a valuable service and to explore their interests. There are other volunteer work experiences available with the Red Cross or community health services. Camp counseling jobs sometimes offer related experiences.

To learn more about career opportunities in the military, visit the Web sites listed at the end of this article.

EMPLOYERS

The U.S. government employs the military. In 2005, 98,482 soldiers were employed in health care occupations: 27,337 individuals served in the air force; 36,823 in the army; 725 in the Coast Guard; and 33,597 in the navy.

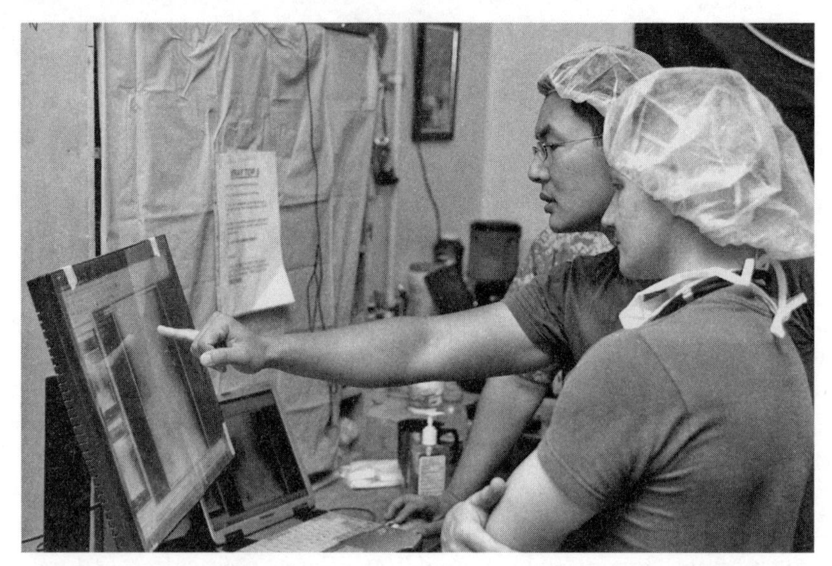

Medical personnel discuss an X ray of an Iraqi police officer at Camp Ramadi, Iraq. The team provided treatment to Iraqi police officers after a vehicle-borne improvised explosive device detonated in Ramadi. *(Sergeant Timothy Stephens, U.S. Army, U.S. Department of Defense)*

STARTING OUT

Talking to a military recruiter should be your first step if you are interested in a career in the military. Visit the Web sites listed at the end of this article to locate a recruiting office near you. To start out in any branch, you will need to pass physical and medical tests, the Armed Services Vocational Aptitude Battery exam, and basic training.

ADVANCEMENT

Each military branch has nine enlisted grades (E-1 through E-9) and 10 officer grades (O-1 through O-10). The higher the number is, the more advanced a person's rank is. The various branches of the military have somewhat different criteria for promoting individuals; in general, however, promotions depend on factors such as length of time served, demonstrated abilities, recommendations, and scores on written exams. Promotions become more and more competitive as people advance in rank. On average, a diligent enlisted person can expect to earn one of the middle noncommissioned or petty officer rankings (E-4 through E-6); some officers can expect to reach lieutenant colonel or commander (O-5). Outstanding individuals may be able to advance beyond these levels.

EARNINGS

The U.S. Congress sets the pay scales for the military after hearing recommendations from the president. The pay for equivalent grades is the same in all services (that is, anyone with a grade of E-4, for example, will have the same basic pay whether in the army, navy, marines, air force, or Coast Guard). In addition to basic pay, personnel who frequently and regularly participate in combat may earn hazardous duty pay. Other special allowances include special duty pay and foreign duty pay. Earnings start relatively low but increase on a fairly regular basis as individuals advance in rank. See the appendix at the end of this book for detailed information on pay scales for the U.S. military. When reviewing earnings, it is important to keep in mind that members of the military receive free housing, food, and health care—items that civilians typically pay for themselves.

Additional benefits for military personnel include uniform allowances, 30 days' paid vacation time per year, and the opportunity to retire after 20 years of service. Generally, those retiring will receive 40 percent of the average of the highest three years of their base pay. This amount rises incrementally, reaching 75 percent of the average of the highest three years of base pay after 30 years of service. All retirement provisions are subject to change, however, and you should verify them as well as current salary information before you enlist. Those who retire after 20 years of service are usually in their 40s and thus have plenty of time, as well as an accumulation of skills, with which to start a second career.

WORK ENVIRONMENT

Health care workers are employed in a variety of settings, including offices, hospitals, clinics, rehabilitation centers, and laboratories. These places are typically clean, well-lit, and climate controlled. Health care workers, such as medical care and service technicians, physicians, and nurses, must exercise meticulous care in their work to avoid risk of exposure to diseases. Some workers may be assigned to work on ships or airplanes. In combat situations, medical care technicians, medical service technicians, physicians, registered nurses, and surgeons may work in mobile field hospitals.

OUTLOOK

Employment in the armed forces is expected to grow about as fast as the average for all occupations through 2014, according to the U.S. Department of Labor. When the economy is good and/or during times of war, more people pursue employment in the civilian workforce, which creates additional opportunities in the military. With the U.S. military involved in several international conflicts, most significantly in Iraq and Afghanistan, demand should continue to be strong for military workers.

Employment for health care workers should be steady as a result of the U.S. military's involvement in Iraq and Afghanistan. Nearly 30,000 soldiers have been injured and more than 4,000 killed during the military conflict in these countries. Opportunities should be strongest for health care workers who provide emergency care, as well as those who help soldiers rehabilitate from injury such as speech therapists and occupational and physical therapy professionals.

The employment outlook in the civilian sector varies by occupation. The U.S. Department of Labor provides the following predictions for professions in the field: dental specialists, medical care technicians, medical service technicians, physical and occupational therapists and specialists, physician assistants, registered nurses, and medical record technicians, much faster than the average; dietitians and optometrists, physicians and surgeons, and health care managers, faster than the average; dentists and speech therapists, about as fast as the average; dental and ophthalmic laboratory technicians, more slowly than the average.

FOR MORE INFORMATION

For more information on physician assistant careers, educational programs, and scholarships, contact

American Academy of Physician Assistants
950 North Washington Street
Alexandria, VA 22314-1552
Tel: 703-836-2272
Email: aapa@aapa.org
http://www.aapa.org

Visit the AACN's Web site to access a list of member schools and to read the online pamphlet, Your Nursing Career: A Look at the Facts.

American Association of Colleges of Nursing (AACN)
One Dupont Circle, Suite 530
Washington, DC 20036-1135
Tel: 202-463-6930
http://www.aacn.nche.edu

For information on state licensing and student resources, contact
American College of Health Care Administrators
300 North Lee Street, Suite 301
Alexandria, VA 22314-2807
Tel: 888-88-ACHCA
http://www.achca.org

For general information on health care management, contact
American College of Healthcare Executives
One North Franklin Street, Suite 1700
Chicago, IL 60606-3529
Tel: 312-424-2800
Email: geninfo@ache.org
http://www.ache.org

For education and career information, contact
American Dental Association
211 East Chicago Avenue
Chicago, IL 60611-2678
Tel: 312-440-2500
http://www.ada.org

The ADA is the single best source of information about careers in dietetics. Its Web site is an excellent resource that provides detailed information and links to other organizations and resources.
American Dietetic Association (ADA)
120 South Riverside Plaza, Suite 2000
Chicago, IL 60606-6995
Tel: 800-877-1600
http://www.eatright.org

For information on earnings, careers in health information management, and accredited programs, contact
American Health Information Management Association
233 North Michigan Avenue, Suite 2150
Chicago, IL 60601-5800
Tel: 312-233-1100
Email: info@ahima.org
http://www.ahima.org

For general information on health care careers, contact
American Medical Association
515 North State Street

Chicago, IL 60610-5453
Tel: 800-621-8335
http://www.ama-assn.org

Visit the AOTA's Web site to find out about accredited occupational therapy programs, career information, and news related to the field.

American Occupational Therapy Association (AOTA)
4720 Montgomery Lane
PO Box 31220
Bethesda, MD 20824-1220
Tel: 301-652-2682
Email: educate@aota.org
http://www.aota.org

For information on optometry careers and accredited educational programs, contact

American Optometric Association
243 North Lindbergh Boulevard
St. Louis, MO 63141-7851
Tel: 800-365-2219
http://www.aoa.org

The APTA offers the brochure Your Career in Physical Therapy, *a directory of accredited schools, and general career information.*

American Physical Therapy Association (APTA)
1111 North Fairfax Street
Alexandria, VA 22314-1488
Tel: 800-999-2782
http://www.apta.org

For educational programs and scholarship information, contact

National Association of Emergency Medical Technicians
PO Box 1400
Clinton, MS 39060-1400
Tel: 800-346-2368
Email: info@naemt.org
http://www.naemt.org

For additional information on nurse assistant careers and training, contact

National Network of Career Nursing Assistants
3577 Easton Road
Norton, OH 44203-5661

Tel: 330-825-9342
Email: cnajeni@aol.com
http://www.cna-network.org

To get information on specific branches of the military, check out this site, which is the home of ArmyTimes.com, NavyTimes.com, and AirForceTimes.com:

Military City
http://www.militarytimes.com

If you're thinking of joining the armed forces, take a look at this site, which guides students and parents through the decision-making process:

Today's Military
http://www.todaysmilitary.com

For information on military careers, contact

United States Air Force Medical Service
http://airforcemedicine.afms.mil

United States Army Medical Department
http://www.armymedicine.army.mil

United States Coast Guard
http://www.gocoastguard.com

United States Navy: Health Care
http://www.navy.com/healthcareopportunities/?campaign=healthcare/medicalservices

Information Technology and Computer Science Occupations

QUICK FACTS

School Subjects
Computer science
Government

Personal Skills
Leadership/management
Technical/scientific

Work Environment
Primarily indoors
Primarily multiple locations

Minimum Education Level
Varies by career specialty

Salary Range
$15,617 to $30,618 to $66,154 (enlisted personnel)
$25,358 to $70,877 to $174,103 (officers)

Outlook
About as fast as the average

DOT
030, 031, 033, 039, 378, 632

GOE
02.01.01, 02.06.01, 02.06.02, 02.07.01

O*NET-SOC
15-1021.00, 15-1031.00, 15-1032.00, 15-1041.00, 15-1051.00, 15-1061.00, 15-1071.00, 15-1081.00, 55-1011.00, 55-2011.00, 55-3011.00, 55-3017.00

OVERVIEW

Military personnel use computers to help complete many tasks such as data entry, accounting, communications, analyzing intelligence information, and controlling weapons systems. *Computer science specialists and officers* design, implement, maintain, repair, and administer the military's computer systems. Opportunities for both enlisted personnel and officers are available in the U.S. Air Force, Army, Coast Guard, Marines, and Navy.

HISTORY

The first substantial developments in modern computer technology took place in the early and mid-twentieth century. After World War II, it was thought that the use of computers would be limited to large government projects, such as the Census, because computers at this time were enormous in size (they easily took up the space of entire warehouses).

Smaller and less expensive computers were made possible due to the introduction of semiconductors. Businesses began using computers in their operations as early as 1954. Within 30 years, computers revolutionized the way people work, play, and shop. Today, computers are everywhere, from businesses of all kinds to government agencies, charitable organizations, and private homes. Over the years, technology has continued to shrink the size of computers and increase computer speed at an unprecedented rate.

Early computers the size of entire rooms were used by the military in the late 1940s and early 1950s. But it was not until the Cold War with Russia that the U.S. government and military began seriously focusing on developing computer technology. In 1958, Russia launched *Sputnik*, the first satellite, and the U.S. government became concerned that it was losing its technological dominance. Soon after, the Advanced Research Projects Agency (ARPA) was created by the U.S. Department of Defense to help the United States regain and maintain a lead in technology. The Department of Defense wanted a comprehensive, indestructible network that could communicate even when under enemy attack. If one of the computer links was destroyed, messages still had to be able to get through to their destination. With this in mind, packet switching was developed in 1968. Within a message (packet) was information about its destination and how to get there so that, if the message hit a point of failure in its route, it could tell where it needed to go next to reach its final destination.

In 1969, using packet switching, ARPA created an internetwork (a network of networks) called ARPANET. There was still a problem, however. How could different types of networks communicate with one another? The U.S. military wanted to link all its networks together worldwide. Despite its inability to link these different networks together, the ARPANET grew. Military and defense contractors, universities, and scientists were rapidly getting connected to this internetwork. After many technological innovations, ARPANET eventually became what we know today as the Internet.

The Internet is not the only computer technology used by the armed forces. For example, the military uses computers to manage its payroll and its personnel, track maintenance schedules for vehicles, facilitate the transportation of personnel and supplies throughout the world, and, most importantly, operate its weapons and intelligence-gathering systems. In short, the military would not be able to function without information technology and computer professionals.

THE JOB

Computer systems specialists plan, develop, and operate all military computer systems used for data storage and processes, communications, and the control of military equipment. They build system networks by installing hardware, software, and peripheral equipment needed for large area networks. They troubleshoot for viruses, compromises in network security, and other problems with system

Rank and Military Branch By Occupation

Job Title	Rank	Military Branches
Computer Systems Officers	Officer	Air Force, Army, Coast Guard, Marines, Navy
Computer Systems Specialists	Enlisted	Air Force, Army, Coast Guard, Marines, Navy

Source: U.S. Department of Defense

requirements. Some specialists serve as computer administrators and provide user training and consultation. Other specialists are involved in data entry and processing.

Computer systems officers identify, develop, and implement the computer needs of the military. They decide what equipment is needed, and evaluate bids from different technological companies and suppliers. They supervise the training of personnel who need to learn how to use new equipment or software. They often coordinate with other department heads to provide computer systems that will be used in upcoming military missions. Computer system officers may work with a team of specialists to design, test, and maintain new software programs and databases.

REQUIREMENTS

High School

In high school you should take any computer programming or computer science courses available. You should also concentrate on math, science, and typing courses.

Postsecondary Training

In the civilian sector, most employers require computer professionals to have at least a bachelor's degree in computer science, computer engineering, computer programming, network administration, information security, or a related field.

In the military, computer systems specialists receive training through classroom instruction, on-the-job experience, and advanced course work. Classes cover a wide range of topics, including network management, programming languages, security issues, and the planning, designing, and testing of computer systems.

As officers in the armed forces, computer systems officers must enter the military with at least a bachelor's degree. Once they join the military, they continue to learn about their profession via classroom instruction. In classes, they learn how to develop computer systems, manage projects, create and manage budgets, and assess the technology needs of their departments.

Other Requirements

Individuals interested in this career should have a solid grasp of math and all aspects of computer science. It helps to be well organized, detail oriented, and able to solve problems in a systematic way. Officers need good management skills as they often are in charge and responsible for a team of specialists.

Visit the U.S. Department of Defense's Web site, http://www.todaysmilitary.com, for more on personal requirements for workers in these careers.

EXPLORING

Try to spend a day with a computer professional in order to experience firsthand what jobs in this industry are like. School guidance counselors can help you arrange such a visit. You can also talk to your high school computer teacher for more information.

In general, you should be intent on learning as much as possible about computers and computer software. You should learn about new developments by reading trade magazines and talking to other computer users. You also can join computer clubs and surf the Internet for information about working in this field.

To learn more about career opportunities in the military, visit the Web sites listed at the end of this article.

EMPLOYERS

The U.S. government employs the military. No specific employment statistics are available for computer workers in the military. Overall, 1.4 million men and women are on active duty and another 1.2 million volunteers serve in the National Guard and Reserve forces.

EARNINGS

The U.S. Congress sets the pay scales for the military after hearing recommendations from the president. The pay for equivalent grades is the same in all services (that is, anyone with a grade of E-4, for example, will have the same basic pay whether in the army, navy, marines, air force, or Coast Guard). In addition to basic pay, personnel who frequently and regularly participate in combat may

Servicemembers Opportunity Colleges Program

Since 1972, military personnel who want to earn an associate's or bachelor's degree while they serve their country have been able to take advantage of the Servicemembers Opportunity Colleges Program. Off-duty student-soldiers in the U.S. Army, Coast Guard, Marine Corps, and Navy can take classes at approximately 1,800 affiliated colleges and universities. These schools are located at or near U.S. military installations, overseas, and on navy ships. Visit http://www.soc.aascu.org for more information.

earn hazardous duty pay. Other special allowances include special duty pay and foreign duty pay. Earnings start relatively low but increase on a fairly regular basis as individuals advance in rank. See the appendix at the end of this book for detailed information on pay scales for the U.S. military. When reviewing earnings, it is important to keep in mind that members of the military receive free housing, food, and health care—items that civilians typically pay for themselves.

Additional benefits for military personnel include uniform allowances, 30 days' paid vacation time per year, and the opportunity to retire after 20 years of service. Generally, those retiring will receive 40 percent of the average of the highest three years of their base pay. This amount rises incrementally, reaching 75 percent of the average of the highest three years of base pay after 30 years of service. All retirement provisions are subject to change, however, and you should verify them as well as current salary information before you enlist. Those who retire after 20 years of service are usually in their 40s and thus have plenty of time, as well as an accumulation of skills, with which to start a second career.

WORK ENVIRONMENT

Computer science specialists and officers work in offices or at computer sites located on military bases, ships, or submarines. Work environments depend on the specialty. Data entry specialists spend the majority of their workday in front of a computer, network specialists may have to travel to remote sites to install new systems, and administrators may spend some time training groups or individuals.

OUTLOOK

Employment in the armed forces is expected to grow about as fast as the average for all occupations through 2014, according to the U.S. Department of Labor. When the economy is good and/or during times of war, more people pursue employment in the civilian workforce, which creates additional opportunities in the military. With the U.S. military involved in several international conflicts, most significantly in Iraq and Afghanistan, demand should continue to be strong for military workers.

Computers and related technology play a key role in the military. Without them, the military would not be able to launch weapons, gather intelligence, manage its massive workforce, or conduct countless other tasks. As a result, employment should be strong for military computer professionals.

In the civilian sector, employment for many computer professionals, such as software engineers, systems administrators, computer systems analysts, network administrators, and database administrators, is expected to increase much faster than average for all occupations through 2014 as technology becomes more sophisticated and organizations continue to adopt and integrate these technologies, making job openings plentiful. Additionally, faster-than-average growth is predicted for computer support specialists.

FOR MORE INFORMATION

For industry information, contact

American Society for Information Science and Technology
1320 Fenwick Lane, Suite 510
Silver Spring, MD 20910-3560
Tel: 301-495-0900
Email: asis@asis.org
http://www.asis.org

For information on internships, student membership, and the student magazine, Crossroads, *contact*

Association for Computing Machinery
1515 Broadway
New York, NY 10036-5701
Tel: 212-869-7440
http://www.acm.org

For information on scholarships, student membership, and the student newsletter, looking.forward, *contact*

IEEE Computer Society
1730 Massachusetts Avenue, NW
Washington, DC 20036-1992
Tel: 202-371-0101
http://www.computer.org

To get information on specific branches of the military, check out this site, which is the home of ArmyTimes.com, NavyTimes.com, AirForceTimes.com, and MarineCorpsTimes.com:

Military City
http://www.militarytimes.com

If you're thinking of joining the armed forces, take a look at this site, which guides students and parents through the decision-making process:

Today's Military
http://www.todaysmilitary.com

For information on military careers, contact

United States Air Force
http://www.airforce.com

United States Army
http://www.goarmy.com

United States Coast Guard
http://www.gocoastguard.com

United States Marine Corps
http://www.marines.com

United States Navy
http://www.navy.com

Intelligence Professionals

OVERVIEW

Intelligence professionals gather information about enemy forces and foreign countries. This information is critical to military operations and national defense.

QUICK FACTS

School Subjects
Foreign language
Government
History

Personal Skills
Communication/ideas
Technical/scientific

Work Environment
Indoors and outdoors
Primarily multiple locations

Minimum Education Level
Varies by career specialty

Salary Range
$15,617 to $30,618 to $66,154 (enlisted personnel)
$25,358 to $70,877 to $174,103 (officers)

Outlook
About as fast as the average

DOT
059

GOE
N/A

O*NET-SOC
N/A

HISTORY

The concept of intelligence gathering comes from ancient times. In a military treatise titled *Ping-fa* (*The Art of War*), written about 400 B.C., the Chinese military philosopher Sun-Tzu mentions the use of secret agents and the importance of good intelligence. Knowledge of an enemy's strengths and weaknesses has always been important to a country's leaders, and so intelligence systems have been used for centuries.

Intelligence gathering has played a major role in contemporary military history. Both the British and the Americans used intelligence operatives during the Revolutionary War in an attempt to gain strategic advantage. The fledgling Continental Congress sent secret agents abroad in 1775, and Benedict Arnold will always be remembered as a spy who switched his allegiance from the colonists to the mother country. Some historians have suggested that World War I resulted from poor intelligence, since none of the countries involved had intended to go to war. With the rapid developments in technology that occurred in the early 20th century, especially in electronics and aeronautics, intelligence operations expanded in the decades after World War I. Operations escalated during World War II, when the U.S. Office of Strategic Services (1929–45) was in operation. The Central Intelligence Agency (CIA), established in

1947, developed out of this office. At that time, the U.S. government believed that espionage was necessary to combat the aggression of the Soviet Union. The CIA continued to expand its activities during the Cold War, when countries were, in essence, engaged in conflict, using intelligence agencies rather than armies.

Some CIA incidents have caused international embarrassment, such as when a Soviet missile shot down a U.S. spy plane that was flying over and photographing Soviet territory in 1960. A scandal involving illegal wiretaps of thousands of Americans who had opposed the Vietnam War caused the CIA to reduce its activities in the late 1970s, although it geared up again during the administration of President Ronald Reagan. (In 2006, the George W. Bush administration had to defend itself against charges that it directed the National Security Agency, another intelligence-gathering organization, to eavesdrop illegally on Americans who were suspected of being linked to terrorism.)

With the fall of Communism and the end of the Cold War, the role of the CIA and intelligence officers changed. Emphasis is now placed on analyzing the constantly changing political and geographic situations in Eastern Europe, Asia, the Middle East, and other parts of the world. Intelligence officers are in demand to provide updated information and insight into how the political and economic circumstances of the world will affect the United States.

THE JOB

The military relies on intelligence professionals to gather information about the size (number of troops, tanks, etc.), location, strength (quality and number of fighter jets, training level of its troops, etc.), and capabilities of enemy forces.

Intelligence specialists gather information by traveling to battle zones to conduct reconnaissance, taking and studying aerial photographs (of foreign bases, ships, aircraft, munitions factories, weapons laboratories, and missile sites), and using electronic monitoring devices such as radar and supersensitive radios. They also study and try to break military codes used by the enemy; prepare intelligence reports, charts, and maps for their superiors; and collect human intelligence, which involves the interviewing or interrogation of prisoners, civilians, or foreign nationals who might be considered threats to national security.

Intelligence officers supervise the work of intelligence specialists. They plan and direct surveillance activities, create plans to intercept foreign communications transmissions, direct the analysis of intelligence that has been gathered, write reports that summarize

Rank and Military Branch by Occupation

Job Title	Rank	Military Branches
Intelligence Officers	Officer	Air Force, Army, Coast Guard, Marines, Navy
Intelligence Specialists	Enlisted	Air Force, Army, Coast Guard, Marines, Navy

Source: U.S. Department of Defense

intelligence activities, brief commanders about intelligence activities and findings, and plan military missions including those that involve electronic warfare.

Specialized intelligence professionals include *technical analysts,* who gather data from satellites, and *cryptographic technicians,* who are experts at coding, decoding, and sending secret messages.

REQUIREMENTS

High School

If you are interested in becoming an intelligence officer or specialist, you can begin preparing in high school by taking courses in English, history, government, journalism, geography, social studies, and foreign languages. You should develop your writing and computer skills as well. Students with the highest grades have the best possibilities for finding employment as intelligence professionals.

Postsecondary Training

In the civilian sector, you must earn at least a bachelor's degree to become an intelligence officer, and an advanced degree is desirable. Specialized skills are also needed for many intelligence roles. The ability to read and speak a foreign language is an asset, as is computer literacy.

In the military, intelligence specialists learn how to do their jobs via classroom instruction, on-the-job training, and advanced course work. During classes, intelligence specialists learn how to gather intelligence, plan aerial and satellite observations, use computer technology to gather intelligence data, analyze data, and prepare intelligence reports, charts, and maps for superiors.

Intelligence officers must enter the military with at least a bachelor's degree. Once they join the military, intelligence officers receive additional specialized training via classroom instruction, on-the-job experience, and advanced course work. Classes teach intelligence officers how to analyze information; conduct ground, air, and sea operations; and use reconnaissance equipment, surveillance equipment, and weapons systems.

Visit the U.S. Department of Defense's Web site, http://www.todaysmilitary.com, for more on military training for those interested in intelligence careers.

Other Requirements

High moral character, discipline, and discretion are essential. Intelligence professionals must have excellent analytical skills, be organized, enjoy reading maps and charts, and have strong oral and written communication skills. They also should be in excellent physical shape.

Visit the U.S. Department of Defense's Web site, http://www.todaysmilitary.com, for more on personal requirements for workers in these careers.

EXPLORING

Opportunities exist for paid and unpaid internships for college undergraduates and graduates at a number of agencies based in the Washington, D.C., area that deal with foreign and defense policy and other matters of interest to intelligence officers. The Federal Bureau of Investigation (FBI), for example, runs the Honors Internship Program during the summer. For more information on this program, visit http://www.fbijobs.gov/231.asp. In addition to this program, the FBI offers several other internship opportunities. Visit http://www.fbijobs.gov for more information.

The CIA runs a Co-Op Program in Washington, D.C., which is open to "highly motivated undergraduates studying a wide variety of fields, including engineering, computer science, mathematics, economics, physical sciences, foreign languages, area studies, business administration, accounting, international relations, finance, logistics, human resources, geography, national security studies, military and foreign affairs, political science and graphic design." Students who are accepted to this program are expected to spend at least three semesters or four quarters on the job prior to graduation. You must apply six to nine months before you are available to work, and you must have at least a 3.0 grade point average. The CIA also has an internship program for students from many backgrounds, including political science and geography majors. Visit https://www.cia.gov/careers for more information on these programs.

Military Glossaries

About.com: U.S. Military Glossary
http://usmilitary.about.com/od/theorderlyroom/l/blglossary.htm

Glossary of Department of Defense Work Force Terms
http://siadapp.dmdc.osd.mil/personnel/NETGLOSS.HTM

U.S. Department of Defense Dictionary of Military and Associated Terms
http://www.dtic.mil/doctrine/jel/doddict

To learn more about career opportunities in the military, visit the Web sites listed at the end of this article.

EMPLOYERS

The U.S. government employs the military. No specific employment statistics are available for military intelligence workers. Overall, 1.4 million men and women are on active duty and another 1.2 million volunteers serve in the National Guard and Reserve forces.

STARTING OUT

Talking to a military recruiter should be your first step if you are interested in a career in the military. Visit the Web sites listed at the end of this article to locate a recruiting office near you. To start out in any branch, you will need to pass physical and medical tests, the Armed Services Vocational Aptitude Battery exam, and basic training.

ADVANCEMENT

Each military branch has nine enlisted grades (E-1 through E-9) and 10 officer grades (O-1 through O-10). The higher the number is, the more advanced a person's rank is. The various branches of the military have somewhat different criteria for promoting individuals; in general, however, promotions depend on factors such as length of time served, demonstrated abilities, recommendations, and scores on written exams. Promotions become more and more competitive as people advance in rank. On average, a diligent enlisted person can expect to earn one of the middle noncommissioned or petty officer rankings (E-4 through E-6); some officers can expect to reach lieutenant colonel or commander (O-5). Outstanding individuals may be able to advance beyond these levels.

EARNINGS

The U.S. Congress sets the pay scales for the military after hearing recommendations from the president. The pay for equivalent grades is the same in all services (that is, anyone with a grade of E-4, for example, will have the same basic pay whether in the army, navy, marines, air force, or Coast Guard). In addition to basic pay, personnel who frequently and regularly participate in combat may earn hazardous duty pay. Other special allowances include special duty pay and foreign duty pay. Earnings start relatively low but increase on a fairly regular basis as individuals advance in rank. See the appendix at the end of this book for detailed information on pay scales for the U.S. military. When reviewing earnings, it is important to keep in mind that members of the military receive free housing, food, and health care—items that civilians typically pay for themselves.

Additional benefits for military personnel include uniform allowances, 30 days' paid vacation time per year, and the opportunity to retire after 20 years of service. Generally, those retiring will receive 40 percent of the average of the highest three years of their base pay. This amount rises incrementally, reaching 75 percent of the average of the highest three years of base pay after 30 years of service. All retirement provisions are subject to change, however, and you should verify them as well as current salary information before you enlist. Those who retire after 20 years of service are usually in their 40s and thus have plenty of time, as well as an accumulation of skills, with which to start a second career.

WORK ENVIRONMENT

Intelligence professionals may find themselves in a laboratory, at a computer station, on a ship, or in a jungle or desert during wartime. Those working on the front lines face danger on a daily basis. Many officers travel often, and travel may include everything from jet planes to small boats to traveling on foot. Even those intelligence professionals who are not working in the field generally work long and erratic hours to meet deadlines for filing reports or completing a mission, especially in times of crisis.

OUTLOOK

Employment in the armed forces is expected to grow about as fast as the average for all occupations through 2014, according to the

U.S. Department of Labor. When the economy is good and/or during times of war, more people pursue employment in the civilian workforce, which creates additional opportunities in the military. With the U.S. military involved in several international conflicts, most significantly in Iraq and Afghanistan, demand should continue to be strong for military workers.

Military intelligence professionals should have excellent employment opportunities. Wars have been won and lost simply on the quality of intelligence; as a result, intelligence workers will continue to be integral members of the military.

Many excellent opportunities are available to intelligence professionals outside of the military, especially with the CIA, Defense Intelligence Agency, FBI, National Security Agency, U.S. Department of Homeland Security, and U.S. Department of State. In general, people with specialized skills or backgrounds in the languages and customs of certain countries will continue to be in high demand.

FOR MORE INFORMATION

For information on the intelligence community and scholarships, contact

Association for Intelligence Officers
6723 Whittier Avenue, Suite 303A
McLean, VA 22101-4582
Tel: 703-790-0320
Email: afio@afio.com
http://www.afio.com

To take a career assessment quiz and for information on career paths, recruitment schedules, and student opportunities, such as internships, visit the CIA's Web site:

Central Intelligence Agency (CIA)
Office of Public Affairs
Washington, DC 20505
Tel: 703-482-0623
http://www.cia.gov/employment

For information on intelligence careers, contact

Defense Intelligence Agency
http://www.dia.mil/employment

Federal Bureau of Investigation
http://www.fbijobs.com

National Security Agency
http://www.nsa.gov

U.S. Department of Homeland Security
http://www.dhs.gov

U.S. Department of State
http://www.careers.state.gov

To get information on specific branches of the military, check out this site, which is the home of ArmyTimes.com, NavyTimes.com, AirForceTimes.com, and MarineCorpsTimes.com:
Military City
http://www.militarytimes.com

For information on intelligence training for members of the military or federal government, contact
National Defense Intelligence College
http://www.dia.mil/college

If you're thinking of joining the armed forces, take a look at this site, which guides students and parents through the decision-making process:
Today's Military
http://www.todaysmilitary.com

For information on military careers, contact
United States Air Force
http://www.airforce.com

United States Army
http://www.goarmy.com

United States Coast Guard
http://www.gocoastguard.com

United States Marine Corps
http://www.marines.com

United States Navy
http://www.navy.com

Interpreters and Translators

OVERVIEW

Interpreters translate spoken passages of a foreign language into another specified language. The job is often designated by the language interpreted, such as Spanish or Arabic. In contrast to interpreters, *translators* focus on written materials that are in a foreign language, such as books, newspapers, magazines, radio and television broadcasts, technical or scientific papers, legal documents, laws, treaties, and decrees.

QUICK FACTS

School Subjects
English
Foreign language
Speech

Personal Skills
Communication/ideas
Helping/teaching

Work Environment
Indoors and outdoors
Primarily multiple locations

Minimum Education Level
Some postsecondary training

Salary Range
$15,617 to $30,618 to $66,154 (enlisted personnel)

Outlook
About as fast as the average

DOT
137

GOE
01.03.01

O*NET-SOC
27-3091.00

HISTORY

Until recently, most people who spoke two languages well enough to interpret and translate did so only on the side, working full time in some other occupation. Interpreting and translating as full-time professions have emerged only recently, partly in response to the need for high-speed communication across the globe. The increasing use of complex diplomacy has also increased demand for full-time translating and interpreting professionals. For many years, diplomacy was practiced largely between just two nations. Rarely did conferences involve more than two languages at one time. The League of Nations, established by the Treaty of Versailles in 1919, established a new pattern of communication. Although the "language of diplomacy" was then considered to be French, diplomatic discussions were carried out in many different languages for the first time. Since the early 1920s, multinational conferences have become commonplace. Trade and educational conferences are now held with participants of many nations in attendance. Responsible for international diplomacy after the League of Nations dissolved, the United Nations now employs many full-time interpreters and translators,

Rank and Military Branch by Occupation

Job Title	Rank	Military Branches
Interpreters and Translators	Enlisted	Air Force, Army, Marines, Navy

Source: U.S. Department of Defense

providing career opportunities for qualified people. In addition, the European Union employs a large number of interpreters.

Understanding an enemy's language and culture has always been a key strategic tool in the art of war. In the American colonies, interpreters and translators served in the military during the French and Indian War, translating and interpreting French and Native American languages. From the Mexican-American War and Indian Wars of the American West, to both World Wars through the war on terrorism today, interpreters and translators have played a key role in helping the military understand its enemies, as well as in communicating with foreign civilians caught up in war.

THE JOB

The military relies on interpreters and translators to convert information that is in a foreign language to English. Our nation's security and the safety of soldiers and U.S. civilian workers in war zones depends on their fast and comprehensive understanding of the written and spoken word.

Military interpreters have many responsibilities. They help intelligence officers interview prisoners of war, civilian informers, and enemy deserters who do not speak English. They work with front-line troops who need to interact with civilians. For example, an interpreter serving in Baghdad, Iraq, might be asked to explain the details of a curfew to residents of a neighborhood that has experienced a large number of attacks on troops by insurgents. Others might work as liaisons between the military and members of the Iraqi government.

While interpreters focus on the spoken word, translators work with written language. They read and translate battle plans, military personnel records, newspapers, magazines, nonfiction and technical works, legal documents, records and reports, speeches, and other written material. Translators generally follow a certain set of procedures in their work. They begin by reading the text, taking careful notes on what they do not understand. To translate questionable passages, they look up words and terms in specialized dictionaries and glossaries. They may also do additional reading on the subject to arrive at a better understanding. Finally, they write translated drafts in the target language.

REQUIREMENTS

High School

If you are interested in becoming an interpreter or translator, you should take a variety of English courses, because most translating work is from a foreign language into English. The study of one or more foreign languages is vital. If you are interested in becoming proficient in one or more of the Romance languages, such as Italian, French, or Spanish, basic courses in Latin will be valuable.

While you should devote as much time as possible to the study of at least one foreign language, other helpful courses include speech, business, cultural studies, humanities, world history, geography, and political science. In fact, any course that emphasizes the written and/or spoken word will be valuable to aspiring interpreters or translators. Finally, courses in typing and word processing are recommended, especially if you want to pursue a career as a translator.

Postsecondary Training

Because interpreters and translators need to be proficient in grammar, have an excellent vocabulary in the chosen language, and have sound knowledge in a wide variety of subjects, civilian employers generally require that applicants have at least a bachelor's degree.

In addition to language and field-specialty skills, you should take college courses that will allow you to develop effective techniques in public speaking, particularly if you're planning to pursue a career as an interpreter. Courses such as speech and debate will improve your diction and confidence as a public speaker.

In the military, interpreters and translators learn how to do their jobs via classroom training, advanced course work, and on-the-job experience. Classes cover the language in which the interpreter is specializing, interrogation methods, and the preparation of reports. Training can last up to 12 months depending on job requirements.

Visit the U.S. Department of Defense's Web site, http://www.todaysmilitary.com, for more on military training for careers in interpreting and translation.

Other Requirements

Interpreters should be able to speak at least two languages fluently, without strong accents. They should be knowledgeable of the foreign

An army soldier and an interpreter question two suspicious Iraqi men in Abd al Hasan, Iraq, during a combat operation. *(Staff Sergeant Dallas Edwards, U.S. Air Force, U.S. Department of Defense)*

language as well as the culture and social norms of the region or country in which it is spoken. For example, an interpreter serving with the military in Afghanistan should be proficient in Pashto and Dari as well as Afghan social and religious customs. Interpreters and translators should read daily newspapers in the languages in which they work to keep current in both developments and usage.

Interpreters must have good hearing, a sharp mind, and a strong, clear, and pleasant voice. They must be able to be precise and quick in their translation. In addition to being flexible and versatile in their work, both interpreters and translators should have self-discipline and patience. Above all, they should have an interest in and love of language.

Visit the U.S. Department of Defense's Web site, http://www.todaysmilitary.com, for more on personal requirements for workers in these careers.

EXPLORING

If you have adequate skills in a foreign language, you might consider traveling in a country in which the language is spoken. If you can converse easily and without a strong accent and can interpret to others who may not understand the language well, you may have what it takes to work as an interpreter or translator.

For any international field, it is important that you familiarize yourself with other cultures. You can even arrange to regularly correspond with a pen pal in a foreign country. You may also want to join a school club that focuses on a particular language, such as the French Club or the Spanish Club. If no such clubs exist, consider forming one. Student clubs can allow you to hone your foreign language speaking and writing skills and learn about other cultures.

Finally, participating on a speech or debate team can allow you to practice your public speaking skills, increase your confidence, and polish your overall appearance by working on eye contact, gestures, facial expressions, tone, and other elements used in public speaking.

To learn more about career opportunities in the military, visit the Web sites listed at the end of this article.

EMPLOYERS

The U.S. government employs the military. No specific employment statistics are available for military interpreters and translators. Overall, 1.4 million men and women are on active duty and another 1.2 million volunteers serve in the National Guard and Reserve forces. Interpreters and translators make up only a small percentage of military workers.

STARTING OUT

The best way to learn more about this career is to talk with a military recruiter. Visit the Web sites listed at the end of this article to locate a recruiting office near you. To start out in any branch, you will need to pass physical and medical tests, the Armed Services Vocational Aptitude Battery exam, and basic training.

ADVANCEMENT

Each military branch has nine enlisted grades (E-1 through E-9) and 10 officer grades (O-1 through O-10). The higher the number is, the more advanced a person's rank is. The various branches of the military have somewhat different criteria for promoting individuals; in general, however, promotions depend on factors such as length of time served, demonstrated abilities, recommendations, and scores on written exams. Promotions become more and more competitive as people advance in rank. On average, a diligent enlisted person can expect to earn one of the middle noncommissioned or petty officer rankings (E-4 through E-6); some officers can expect to reach lieutenant colonel or commander (O-5). Outstanding individuals may be able to advance beyond these levels.

EARNINGS

The U.S. Congress sets the pay scales for the military after hearing recommendations from the president. The pay for equivalent grades is the same in all services (that is, anyone with a grade of E-4, for example, will have the same basic pay whether in the army, navy, marines, air force, or Coast Guard). In addition to basic pay, personnel who frequently and regularly participate in combat may earn hazardous duty pay. Other special allowances include special duty pay and foreign duty pay. Earnings start relatively low but increase on a fairly regular basis as individuals advance in rank. See the appendix at the end of this book for detailed information on pay scales for the U.S. military. When reviewing earnings, it is important to keep in mind that members of the military receive free housing, food, and health care—items that civilians typically pay for themselves.

Additional benefits for military personnel include uniform allowances, 30 days' paid vacation time per year, and the opportunity to retire after 20 years of service. Generally, those retiring will receive 40 percent of the average of the highest three years of their base pay. This amount rises incrementally, reaching 75 percent of the average of the highest three years of base pay after 30 years of service. All retirement provisions are subject to change, however, and you should verify them as well as current salary information before you enlist. Those who retire after 20 years of service are usually in their 40s and thus have plenty of time, as well as an accumulation of skills, with which to start a second career.

WORK ENVIRONMENT

Military interpreters work on military bases, in war zones, aboard ships, and in airplanes. They work in a wide variety of circumstances and conditions. As a result, most do not have typical nine-to-five schedules.

While both interpreting and translating require flexibility and versatility, interpreters in particular, especially those who work in war zones or military prisons, may experience considerable stress and fatigue. Working in potentially dangerous settings and knowing that a great deal depends upon their absolute accuracy in interpretation can be a weighty responsibility.

OUTLOOK

Employment in the armed forces is expected to grow about as fast as the average for all occupations through 2014, according to the

U.S. Department of Labor. When the economy is good and/or during times of war, more people pursue employment in the civilian workforce, which creates additional opportunities in the military. With the U.S. military involved in several international conflicts, most significantly in Iraq and Afghanistan, demand should continue to be strong for military workers.

The U.S. military is presently experiencing a shortage of interpreters and translators who are proficient in Arabic, Kurdish, Persian, Dari, and Pashto—languages that are spoken in Iraq, Iran, or Afghanistan. Interpreters and translators who are skilled in one or more of these languages will have excellent employment opportunities.

Employment opportunities for interpreters and translators in the civilian sector and in government are expected to grow faster than the average for all occupations through 2014, according to the U.S. Department of Labor. However, competition for available positions will be fierce.

FOR MORE INFORMATION

For information on careers in literary translation, contact

American Literary Translators Association
University of Texas-Dallas
Box 830688, Mail Station JO51
Richardson, TX 75083-0688
http://www.literarytranslators.org

For more on the translating and interpreting professions, contact

American Translators Association
225 Reinekers Lane, Suite 590
Alexandria, VA 22314-2875
Tel: 703-683-6100
Email: ata@atanet.org
http://www.atanet.org

For more information on court interpreting, contact

National Association of Judiciary Interpreters and Translators
1707 L Street, NW, Suite 507
Washington, DC 20036-4201
Tel: 202-293-0342
Email: headquarters@najit.org
http://www.najit.org

For information on union membership for freelance interpreters and translators, contact

Translators and Interpreters Guild
962 Wayne Avenue, Suite 500
Silver Spring, MD 20910-4432
Tel: 800-992-0367
Email: info@ttig.org
http://www.ttig.org

To get information on specific branches of the military, check out this site, which is the home of ArmyTimes.com, NavyTimes.com, AirForceTimes.com, and MarineCorpsTimes.com:

Military City
http://www.militarytimes.com

If you're thinking of joining the armed forces, take a look at this site, which guides students and parents through the decision-making process:

Today's Military
http://www.todaysmilitary.com

For information on military careers, contact

United States Air Force
http://www.airforce.com

United States Army
http://www.goarmy.com

United States Coast Guard
http://www.gocoastguard.com

United States Marine Corps
http://www.marines.com

United States Navy
http://www.navy.com

Law Enforcement, Security, and Protective Services Occupations

OVERVIEW

Law enforcement, security, and protective services workers provide police, fire, and safety services to members of the military, their families, and civilian personnel who live and work on military bases, ships, and aircraft. These professionals also work in war zones to protect military personnel and civilians. Opportunities are available for both enlisted personnel and officers in the U.S. Air Force, Army, Coast Guard, Marines, and Navy. In 2005, 80,579 soldiers were employed in law enforcement and protective services occupations.

QUICK FACTS

School Subjects
Physical education
Psychology

Personal Skills
Leadership/management

Work Environment
Indoors and outdoors
Primarily multiple locations

Minimum Education Level
Varies by career specialty

Salary Range
$15,617 to $30,618 to $66,154 (enlisted personnel)
$25,358 to $70,877 to $174,103 (officers)

Outlook
About as fast as the average

DOT
250, 375

GOE
04.01.01, 04.03.01, 04.03.03

O*NET-SOC
33-2011.00, 33-3021.01, 33-3021.02, 33-3051.00, 33-3051.01

HISTORY

Civilization would not exist without fire, but this essential tool can often turn destructive and deadly. For centuries, people have fought to protect their lives and property from fire. The first permanent fire-fighting company in America was formed by Benjamin Franklin in Philadelphia in 1736. New York established its own fire company in 1737, and the practice spread throughout the other colonies. At the same time, volunteer fire brigades supplemented these professional firefighters. The growth of U.S. cities during the 19th century led to an increased need for professional firefighters and better equipment. Many cities suffered devastating fires. Crowded conditions, poor building techniques and materials, the lack of a sufficient water supply, and the absence of coordinated, citywide fire services meant that even a small fire could have terrible consequences. By the turn of the century, almost every large city in the United States had organized professional, paid fire departments, with steam-powered fire engines and a system of fire hydrants to provide an adequate supply of water wherever a fire occurred. Because of an increased knowledge of fire science, fire fighting has evolved from a relatively low-skilled but noble occupation to a highly specialized, technical profession that has expanded considerably in scope to keep pace with the staggering losses of life and property from fire. Today, more than 30,000 organized fire departments are in operation across the United States.

Public safety, and the rules that go with it, has been a concern for a long time. In the earliest societies it was clear that people would run wild unless certain rules of conduct were created. Some laws evolved from the common agreement of the group's members, while others were handed down by the group's leaders. Soon after the establishment of rules and laws, methods of enforcement developed. For a long time, enforcement simply meant punishment. Those who broke the laws were often ostracized or exiled from the group, subjected to corporal punishment, tortured, maimed, or even killed. Enforcement of the law was usually left up to the society's leaders or rulers, often through the soldiers who served in their armies. Often these armies also collected taxes, which were used to maintain the army and sometimes to line the ruler's pockets. Eventually, more organized methods of public safety were developed. The first modern police force was formed in 1829 in London. Cities in the United States organized police forces as well, beginning with New York in 1844. As the United States stretched across the continent, many states created state police forces to work with police forces in the cities and towns. Interstate crimes were placed under federal authority, and various agencies, including the U.S. Marshals Office, Federal Bureau of Investigation, Secret Service, Internal Revenue Service, and the Customs Service, were formed to enforce laws across various jurisdictions.

Throughout military history workers have protected military personnel, their families, civilian workers, and people in war zones from fire, crime, and other dangerous situations. These professionals play a key role in preserving order and safety on military bases, during humanitarian missions, and in war zones throughout the world.

THE JOB

All military bases have their own fire departments, which keep people and property safe from the serious threat of fire. Firefighters provide this protection by fighting fires to prevent property damage and by rescuing people trapped or injured by fires or other accidents. Through inspections and safety education, firefighters also work to prevent fires

Rank and Military Branch by Occupation

Job Title	Rank	Military Branches
Firefighters	Enlisted	Air Force, Army, Coast Guard, Marines, Navy
Law Enforcement and Security Officers	Officer	Air Force, Army, Coast Guard, Marines, Navy
Law Enforcement and Security Specialists	Enlisted	Air Force, Army, Coast Guard, Marines, Navy

Source: U.S. Department of Defense

and unsafe conditions that could result in dangerous, life-threatening situations. They assist in many types of emergencies and disasters.

Law enforcement and security specialists and officers enforce military law on military bases, in coastal wars, and in war zones or other areas where the U.S. military is present. Law enforcement and security specialists have many of the same duties as civilian police professionals. They investigate crimes and activities related to espionage, terrorism, and treason; interview potential witnesses and suspects; patrol military bases and other areas on foot, by car, or boat; direct traffic; respond to riots or other emergencies; and guard inmates in correctional facilities and other military installations.

Law enforcement and security officers manage law enforcement and security specialists; plan and develop programs that reduce crime; plan and direct investigations of suspected espionage, terrorism, and treason; oversee military prisons; and develop and implement security plans for military bases and other facilities.

REQUIREMENTS

High School

If you are interested in becoming a firefighter, you should take science classes such as anatomy, physics, and biology.

If you are interested in pursuing a career in law enforcement, you will find the subjects of psychology, sociology, speech, English, law, mathematics, U.S. government and history, chemistry, and physics most helpful. Knowledge of a foreign language is especially helpful, and bilingual professionals are often in great demand. If specialized and advanced positions in law enforcement interest you, pursue studies leading to college programs in criminology, criminal law, criminal psychology, or related areas.

Because physical stamina is very important in law enforcement and protective services careers, sports and physical education are also valuable.

Postsecondary Training

Educational requirements in the civilian sector vary by career. Many firefighters prepare for the field by studying fire science technology at two- and four-year colleges. Beginning firefighters may receive six to 12 or more weeks of intensive training, either as on-the-job training or through formal fire department training schools.

Many police departments now require a two- or four-year degree, especially for more specialized areas of police work. There are more than 800 junior colleges and universities offering two- and four-year degree programs in law enforcement, police science, and administration of justice. Newly recruited police officers must pass a special training program. After training, they are usually placed on a probationary period lasting from three to six months. In small towns and communities, a new officer may get his or her training on the job by working with an experienced officer. Inexperienced officers are never sent out on patrol alone but are always accompanied by veteran officers.

In the military, enlisted personnel (firefighters and law enforcement and security specialists) train via classroom instruction and on-the-job experience. Course work varies by specialty. For example, firefighters learn about the types of fires, firefighting and rescue procedures, first aid, and the care of equipment. They also get hands-on experience fighting fires. Law enforcement and security specialists receive classroom training regarding military and civil laws, investigative techniques, prisoner control and discipline, use of firearms and hand-to-hand defense techniques, and traffic-control techniques.

As officers in the military, law enforcement and security officers must enter the armed forces with at least a bachelor's degree. Once they join the military, these professionals receive additional specialized training via classroom instruction. Typical courses focus on law enforcement administration, managing security issues, military law, and methods of investigation.

Visit the U.S. Department of Defense's Web site, http://www.todaysmilitary.com, for more on military training in law enforcement, security, and protective services careers.

Other Requirements

Firefighters need sound judgment, mental alertness, and the ability to reason and think logically in situations demanding courage and bravery. Law enforcement professionals must be able to think clearly and

A military policeman uses a mirror to inspect the underside of a vehicle at a checkpoint. *(MCSN Brian Goodwin, U.S. Navy, U.S. Department of Defense)*

logically during emergency situations, have a strong degree of emotional control, and be capable of detaching themselves from incidents. Both firefighters and law enforcement workers must be in top physical condition and be willing to work in sometimes dangerous situations.

Visit the U.S. Department of Defense's Web site, http://www.todaysmilitary.com, for more on personal requirements for careers in law enforcement and protective services.

EXPLORING

To learn more about a career as a firefighter, talk with local firefighters. You may also be able to get permission to sit in on some of the formal training classes for firefighters offered by city fire departments. In some cases, depending on the size and regulations of the town or city department, you may be able to gain experience by working as a volunteer firefighter. Courses in lifesaving and first aid will offer you experience in these aspects of the firefighter's job. You can explore these areas through community training courses and the training offered by the Boy Scouts of America and Girl Scouts of America or the American Red Cross.

A good way to explore police work is to talk with various law enforcement officers. Most departments have community outreach programs and many have recruiting programs as well. You may also wish to visit colleges offering programs in police work or write for information on their training programs.

Many departments also offer explorer and cadet programs for high-school age students who are interested in firefighting or police work.

Others ways to learn about any of the jobs listed in this article include reading books and magazines about firefighting and law enforcement, visiting Web sites of professional associations, and arranging tours of fire or police stations.

To learn more about career opportunities in the military, visit the Web sites listed at the end of this article.

EMPLOYERS

The U.S. government employs the military. In 2005, 80,579 soldiers were employed in law enforcement and protective services occupations: 33,126 individuals served in the air force; 25,507 in the army; 2,799 in the coast guard; 6,042 in the marines; and 13,105 in the navy.

STARTING OUT

Contact a military recruiter for information on entering the armed forces. Visit the Web sites listed at the end of this article to locate a recruiting office near you. To start out in any branch, you will need to pass physical and medical tests, the Armed Services Vocational Aptitude Battery exam, and basic training.

ADVANCEMENT

Each military branch has nine enlisted grades (E-1 through E-9) and 10 officer grades (O-1 through O-10). The higher the number is, the more advanced a person's rank is. The various branches of the military have somewhat different criteria for promoting individuals; in general, however, promotions depend on factors such as length of time served, demonstrated abilities, recommendations, and scores on written exams. Promotions become more and more competitive as people advance in rank. On average, a diligent enlisted person can expect to earn one of the middle noncommissioned or petty officer rankings (E-4 through E-6); some officers can expect to reach lieutenant colonel or commander (O-5). Outstanding individuals may be able to advance beyond these levels.

EARNINGS

The U.S. Congress sets the pay scales for the military after hearing recommendations from the president. The pay for equivalent grades is the same in all services (that is, anyone with a grade of E-4, for example, will have the same basic pay whether in the army, navy, marines, air force, or coast guard). In addition to basic pay, personnel who frequently and regularly participate in combat may earn hazardous duty pay. Other special allowances include special duty pay and foreign duty pay. Earnings start relatively low but increase on a fairly regular basis as individuals advance in rank. See the appendix at the end of this book for detailed information on pay scales for the U.S. military. When reviewing earnings, it is important to keep in mind that members of the military receive free housing, food, and health care—items that civilians typically pay for themselves.

Additional benefits for military personnel include uniform allowances, 30 days' paid vacation time per year, and the opportunity to retire after 20 years of service. Generally, those retiring will receive 40 percent of the average of the highest three years of their base pay. This amount rises incrementally, reaching 75 percent of the average of the highest three years of base pay after 30 years of service. All retirement provisions are subject to change, however, and you should verify them as well as current salary information before you enlist. Those who retire after 20 years of service are usually in their 40s and thus have plenty of time, as well as an accumulation of skills, with which to start a second career.

WORK ENVIRONMENT

The work of firefighters can often be exciting; the job, however, is one of grave responsibilities. Someone's life or death often hangs in the balance. Working conditions are frequently dangerous and involve risking one's life in many situations. Floors, walls, or even entire buildings can cave in on firefighters as they work to save lives and property in raging fires. Exposure to smoke, fumes, chemicals, and gases can end a firefighter's life or cause permanent injury.

Law enforcement and security officers spend a considerable amount of time in offices planning and directing security activities. They work outdoors when overseeing investigations and inspecting security systems and facilities.

Law enforcement and security specialists work under many different types of circumstances. Much of their work is performed outdoors, as they conduct investigations, staff security checkpoints, or patrol facilities, including shipyards, airstrips, and other military installations. The work demands constant mental and physical alertness as well as great physical strength and stamina.

These occupations are considered dangerous. Some workers are killed or wounded while performing their duties.

OUTLOOK

Employment in the armed forces is expected to grow about as fast as the average for all occupations through 2014, according to the U.S. Department of Labor. When the economy is good and/or during times of war, more people pursue employment in the civilian workforce, which creates additional opportunities in the military. With the U.S. military currently involved in several international conflicts, most significantly in Iraq and Afghanistan, demand should continue to be strong for military workers. Military firefighters and law enforcement professionals should have excellent employment opportunities.

The employment outlook in the civilian sector varies by occupation. Employment of firefighters is expected to grow faster than the average for all occupations through 2014, according to the U.S. Department of Labor. Employment of police officers, security officers, and security guards is expected to increase about as fast as the average.

FOR MORE INFORMATION

The IAFF's Web site has a virtual academy with information on scholarships for postsecondary education.

International Association of Fire Fighters (IAFF)
1750 New York Avenue, NW
Washington, DC 20006-5301
Tel: 202-737-8484
http://www.iaff.org

The National Association of Police Organizations is a coalition of police unions and associations that work to advance the interests of law enforcement officers through legislation, political action, and education.

National Association of Police Organizations
317 South Patrick Street
Alexandria, VA 22314-3501
Tel: 703-549-0775
Email: info@napo.org
http://www.napo.org

To get information on specific branches of the military, check out this site, which is the home of ArmyTimes.com, NavyTimes.com, AirForceTimes.com, and MarineCorpsTimes.com:

Military City
http://www.militarytimes.com

If you're thinking of joining the armed forces, take a look at this site, which guides students and parents through the decision-making process:

Today's Military
http://www.todaysmilitary.com

For information on careers in the military, contact

United States Air Force
http://www.airforce.com

United States Army
http://www.goarmy.com

United States Coast Guard
http://www.gocoastguard.com

United States Marine Corps
http://www.marines.com

United States Navy
http://www.navy.com

For information on work in military policing, visit

Military Police Association
http://www.militarypoliceassn.com

United States Air Force Security Forces
http://afsf.lackland.af.mil

United States Army Military Police School
http://www.wood.army.mil/usamps

Legal Professionals and Support Occupations

QUICK FACTS

School Subjects
English
Government
Speech

Personal Skills
Communication/ideas
Leadership/management

Work Environment
Primarily indoors
Primarily multiple locations

Minimum Education Level
Varies by career specialty

Salary Range
$15,617 to $30,618 to $66,154 (enlisted personnel)
$25,358 to $70,877 to $174,103 (officers)

Outlook
About as fast as the average

DOT
110, 111, 201, 202

GOE
04.02.01, 04.02.02, 09.02.02

O*NET-SOC
23-1011.00, 23-1021.00, 23-1023.00, 23-2091.00, 43-6012.00

OVERVIEW

Military *legal professionals and support professionals* work in many areas of the law, including administrative, criminal, environmental, international, maritime, and tort law. They provide legal services to military personnel, retirees, and their families. Opportunities are available for both enlisted personnel and officers in the U.S. Air Force, Army, Coast Guard, Marines, and Navy.

HISTORY

Throughout history, societies have established systems of law to govern people. One of the earliest known codes of law is the Code of Hammurabi developed about 1800 B.C. by the Babylonians. Roughly 400 years later, Moses was given and then introduced the Ten Commandments, which have become the foundation of Judeo-Christian ethics and the basis of our current legal system.

The ancient Greeks and Romans set up the first schools of law for young boys to learn the many skills involved in pleading a case. To be an eloquent speaker was a great advantage. The Greeks especially focused their training on thinking logically, another important part of debating and proving matters of the law.

In America's early days, the settlers lived under the English Common law that they brought with them to the New World. This common law began under Henry II in England and set up standard punishments for certain crimes. This common law was later modified by the Articles of Confederation and then the Constitution. In areas of the United States that were originally controlled by the Spanish, such as California and Texas, traces of

Spanish law still exist. Similarly, because Louisiana was controlled by the French, this state still operates a more French than English legal system. But in all cases, the laws of other countries and other times were adapted by legislatures to fit the changing needs and customs of American society. Our legal system today is called statutory because lawmakers enact statutes to govern us.

Several fields of law have been practiced in the past and continue to be practiced today. The two most well-known fields are criminal law and civil law. Criminal law focuses on crime, that is, acts committed in violation of a law. This field of law is concerned with an individual's relation to society; in fact, criminal acts are considered offenses against all members of a society even if the acts were committed against only one person. Anyone who breaks criminal laws may be punished with prison time or fines. Civil law focuses on relationships between individuals. Most civil law addresses written contract, wrongs against individuals (or torts), and property issues. Breakers of civil law cannot be imprisoned or fined as a result, but they may have to pay money as a result of the decision of the court. Other fields of law include administrative law, constitutional law, tax law, maritime law, and labor law.

In the U.S. military, laws have always existed to maintain order and create a clear-cut system when members of the military are accused of committing crimes against other military personnel, U.S. civilians, or those in war zones. Early military law was roughly based on British law, but as the law became more complicated, it became clear that each branch needed legal experts to help judges interpret existing laws. In response, a Judge Advocate General's Corps was set up by the army in 1775 and followed by the navy (1865), Coast Guard (1906), air force (1948), and Marine Corps (1966).

Despite the creation of these agencies, military legal systems largely existed to "enforce discipline, not justice," until the 1950s, according to *U.S. News & World Report*. Changes were gradually made to the existing laws to encourage more fairness in the legal process. During World War I, three levels of courts-martial were instituted depending on the seriousness of the charges. Events that occurred during World War II created a sea change regarding military law. According to *U.S. News & World Report*, the war produced 2 million courts-martial (or nearly 13 percent of all those who served), with more than 100 servicemen executed and 45,000 jailed. Public outcry caused the military to rethink its approach and President Harry Truman signed the Uniform Code of Military Justice into law on May 31, 1951. The code created a system of checks and balances on military courts that include an appeals system.

Rank and Military Branch by Occupation

Job Title	Rank	Military Branches
Lawyers	Officer	Air Force, Army, Coast Guard, Marines, Navy
Judges	Officer	Air Force, Army, Coast Guard, Marines, Navy
Legal Specialists	Enlisted	Air Force, Army, Coast Guard, Marines, Navy
Court Reporters	Enlisted	Air Force, Army, Coast Guard, Marines, Navy

Source: U.S. Department of Defense

THE JOB

Lawyers provide legal advice and represent clients in court when necessary. In the military, they are employed by the Judge Advocate General's Corps. Military lawyers serve in three ways in the military judicial system: as prosecutors, advocates, and advisors. Prosecuting attorneys represent the military in court proceedings. They gather and analyze evidence and review legal material in order to present evidence against the defendant. As advocates, they represent the rights of military personnel in trials and depositions or in front of other government bodies. As advisors, lawyers prepare legal documents such as wills and powers of attorney for military personnel and provide advice about government real estate, patents, commercial contracts, and trademarks.

Judges preside over military courts. They apply military law to soldiers and oversee court proceedings according to the established law.

Legal specialists assist military lawyers and judges by performing administrative and clerical duties in a law office or court. They write legal correspondence, prepare legal documents, conduct research about court decisions and military regulations, and answer incoming calls and emails. Legal specialists maintain files detailing records of investigations, hearings, courts-martial, and courts of inquiry; take notes during meetings or hearings; maintain law libraries; and assume all other general secretarial duties.

Court reporters record every word at hearings, trials, tribunals, depositions, and other legal proceedings by using a stenotype machine to take shorthand notes. Most court reporters transcribe the notes of the proceedings by using computer-aided transcription systems that print out regular, legible copies of the proceedings.

Court reporters must also edit and proofread the final transcript and create the official transcript of the trial or other legal proceeding.

REQUIREMENTS

High School

If you are considering a career as a lawyer or judge, courses such as government, history, social studies, and economics provide a solid background for entering college-level courses. Speech courses will help you to build strong communication skills necessary for these professions. Also take advantage of any computer-related classes or experience you can get, because lawyers and judges often use technology to research and interpret the law, from surfing the Internet to searching legal databases.

If you are interested in becoming a court reporter or legal secretary, take as many high-level classes in English as you can and get a firm handle on grammar and spelling. Take typing classes and computer classes to get a foundation in using computers and a head start in keyboarding skills. Classes in government and business will be helpful as well.

Postsecondary Training

Educational requirements in the civilian sector vary by career. For example, lawyers and judges must have a law degree to practice in their profession. A high school diploma, a college degree, and three years of law school are minimum requirements for a law degree. Court reporters are required to complete a specialized training program in shorthand reporting. These programs usually last between two and four years and include instruction on how to enter at least 225 words a minute on a stenotype machine. Many legal secretaries get their training through established one- or two-year legal secretary programs. These programs are available at most business, vocational, and junior colleges.

In the military, enlisted personnel (legal specialists and court reporters) receive their training via classroom instruction. They learn legal terminology and about the military judicial system, how to conduct research and prepare legal documents, and, in the case of court reporters, how to perform high-speed legal transcription.

As officers in the military, lawyers and judges must enter the armed forces with a law degree. Once they join the military, they continue their training by taking basic and advanced classes, as well as honing their skills on the job. Typical classes focuses on the application of the Uniform Code of Military Justice, military trial procedures, and court-martial advocacy techniques.

Visit the U.S. Department of Defense's Web site, http://www.todaysmilitary.com, for more on military training for careers in law and legal support.

Other Requirements

Both lawyers and judges have to be effective communicators, work well with people, be attentive to detail, and be able to find creative solutions to problems, such as complex court cases. Court reporters should be familiar with a wide range of medical and legal terms and must be assertive enough to ask for clarification if a term or phrase goes by without the reporter understanding it. They must be as unbiased as possible and accurately record what is said, not what they believe to be true. Patience and perfectionism are vital characteristics, as is the ability to work closely with judges and other court officials. Legal specialists must have knowledge of legal terminology, be able to prioritize and balance different tasks, and have good organization skills.

Visit the U.S. Department of Defense's Web site, http://www.todaysmilitary.com, for more on personal requirements for workers in these careers.

EXPLORING

If you think a career as a lawyer, judge, or court reporter might be right up your alley, there are several ways you can find out more about it before making that final decision. First, sit in on a trial or two at your local or state courthouse. Try to focus mainly on the judge, lawyer, and court reporter and take note of what they do. Write down questions you have and terms or actions you don't understand. Then, talk to your guidance counselor and ask for help in setting up a telephone or in-person interview with a judge, lawyer, or court reporter. Ask questions and get the scoop on what those careers are really all about. Also, talk to your guidance counselor or political science teacher about starting or joining a job shadowing program. Job shadowing programs allow you to follow a person in a certain career for a day or two to get an idea of what goes on in a typical day. You may even be invited to help out with a few minor duties. You can also search the World Wide Web for general information about lawyers, judges, and court reporters and current court cases. Read court transcripts and summary opinions written by judges on issues of importance today. After you've done some research and talked to a lawyer, judge, or court reporter and you still think you are destined for a career in law, try to get a part-time job in a law office or a court. Ask your guidance counselor for help.

Books to Read

Beyer, Rick. *The Greatest War Stories Never Told: 100 Tales from Military History to Astonish, Bewilder, and Stupefy.* New York: Collins, 2005.

Bonn, Keith E. *Army Officer's Guide.* 50th ed. Mechanicsburg, Pa.: Stackpole Books, 2005.

Ostrow, Scott A. *Guide to Joining the Military.* 2d ed. Lawrenceville, N.J.: ARCO/Peterson's, 2003.

Paradis, Adrian A. *Opportunities in Military Careers.* Rev. ed. New York: McGraw-Hill, 2005.

Samet, Elizabeth D. *Soldier's Heart: Reading Literature Through Peace and War at West Point.* New York: Farrar, Straus and Giroux, 2007.

U.S. Department of Defense. *America's Top Military Careers: Official Guide to Occupations in the Armed Forces.* 4th ed. Indianapolis, Ind.: JIST Publishing, 2003.

To learn more about career opportunities in the military, visit the Web sites listed at the end of this article.

EMPLOYERS

The U.S. government employs the military. Today 1.4 million men and women are on active duty and another 1.2 million volunteers serve in the National Guard and Reserve forces. Legal professionals make up only a small percentage of military workers.

STARTING OUT

If you are considering a career in the military, your first step should be to contact a military recruiter to learn more about your career options. Visit the Web sites listed at the end of this article to locate a recruiting office near you. To start out in any branch, you will need to pass physical and medical tests, the Armed Services Vocational Aptitude Battery exam, and basic training.

ADVANCEMENT

Each military branch has nine enlisted grades (E-1 through E-9) and 10 officer grades (O-1 through O-10). The higher the number is, the more advanced a person's rank is. The various branches of the military have somewhat different criteria for promoting individuals; in general, however, promotions depend on factors such as length of time served, demonstrated abilities, recommendations, and scores on written exams. Promotions become more and more competitive as people advance in rank. On average, a diligent enlisted person can expect to earn one of the middle noncommissioned or petty officer rankings (E-4 through E-6); some officers can expect to reach lieutenant colonel or commander (O-5). Outstanding individuals may be able to advance beyond these levels.

EARNINGS

The U.S. Congress sets the pay scales for the military after hearing recommendations from the president. The pay for equivalent grades is the same in all services (that is, anyone with a grade of E-4, for example, will have the same basic pay whether in the army, navy, marines, air force, or Coast Guard). In addition to basic pay, personnel who frequently and regularly participate in combat may earn hazardous duty pay. Other special allowances include special duty pay and foreign duty pay. Earnings start relatively low but increase on a fairly regular basis as individuals advance in rank. See the appendix at the end of this book for detailed information on pay scales for the U.S. military. When reviewing earnings, it is important to keep in mind that members of the military receive free housing, food, and health care—items that civilians typically pay for themselves.

Additional benefits for military personnel include uniform allowances, 30 days' paid vacation time per year, and the opportunity to retire after 20 years of service. Generally, those retiring will receive 40 percent of the average of the highest three years of their base pay. This amount rises incrementally, reaching 75 percent of the average of the highest three years of base pay after 30 years of service. All retirement provisions are subject to change, however, and you should verify them as well as current salary information before you enlist. Those who retire after 20 years of service are usually in their 40s and thus have plenty of time, as well as an accumulation of skills, with which to start a second career.

WORK ENVIRONMENT

Lawyers, judges, legal specialists, and court reporters work in offices and courtrooms on land and aboard ships. Offices are usually pleasant, although busy, places to work. Lawyers also spend significant amounts of time in law libraries or record rooms and sometimes in the jail cells of clients or prospective witnesses. Many lawyers never work in a courtroom.

OUTLOOK

Employment in the armed forces is expected to grow about as fast as the average for all occupations through 2014, according to the U.S. Department of Labor. When the economy is good and/or during times of war, more people pursue employment in the civilian workforce, which creates additional opportunities in the military. With the U.S. military currently involved in several international conflicts, most significantly in Iraq and Afghanistan, demand should continue to be strong for military workers, including legal professionals.

In the civilian sector, employment for lawyers, judges, court reporters, and legal secretaries is expected to grow about as fast as the average for all occupations through 2014, according to the U.S. Department of Labor.

FOR MORE INFORMATION

For information on careers, contact

American Bar Association
321 North Clark Street
Chicago, IL 60610-4714
Tel: 800-285-2221
http://www.abanet.org

For information on legal secretary careers, contact

National Association of Legal Secretaries
8159 East 41st Street
Tulsa, OK 74145-3312
Tel: 918-582-5188
Email: info@nals.org
http://www.nals.org

For information on court reporting careers, contact

National Court Reporters Association
8224 Old Courthouse Road
Vienna, VA 22182-3808
Tel: 800-272-6272
Email: msic@ncrahq.org
http://www.ncraonline.org

For information on military law, visit the following Web sites:

Judge Advocate General's Corps: U.S. Air Force
http://hqja.jag.af.mil

Judge Advocate General's Corps: U.S. Army
http://www.goarmy.com/jag

U.S. Coast Guard Legal Division
http://www.uscg.mil/legal

Judge Advocate General's Corps: U.S. Marine Corps
http://sja.hqmc.usmc.mil

Judge Advocate General's Corps: U.S. Navy
http://www.jag.navy.mil

To get information on specific branches of the military, check out this site, which is the home of ArmyTimes.com, NavyTimes.com, AirForceTimes.com, and MarineCorpsTimes.com:
Military City
http://www.militarytimes.com

If you're thinking of joining the armed forces, take a look at this site, which guides students and parents through the decision-making process:
Today's Military
http://www.todaysmilitary.com

INTERVIEW

Erica Austin is a military lawyer who is currently assigned to the 944th Fighter Wing, Luke Air Force Base, Arizona. She discussed her career with the editors of Careers in Focus: Armed Forces.

Q. Can you please tell us about yourself?

A. I am currently in the air force reserves. I am a judge advocate or, in other words, an air force attorney. Until August 2007 I was an active duty Air Force member also serving in the Judge Advocate General (JAG) Corps. I have been an officer in the air force for 13 years. I have served in Texas; Rome, Italy; Lillehammer, Norway; Georgia; Alaska; Colorado; Korea; Japan; Guam; Arizona; and Hawaii (some short-term assignments; others long-term assignments).

Q. What made you want to enter the military?

A. I wanted to help people. I was in the Reserve Officer Training Corps while attending the University of Illinois and, somewhere near the end of my time there, I started to think about what I could do in the air force that would be fulfilling, challenging, and rewarding. I grew up around lawyers (my mother worked for lawyers since she was 15 years old), and I had met some along the way that I knew I did not want to emulate. I decided to go to law school and become an air force attorney, which is what I did—and I have no regrets. The military is an environment based on integrity and trust. Without those personal qualities, people fail in the military. I have always thrived in positive environments in which I am surrounded by people who believe and think similarly to me. The air force is a place where people sacrifice their time and selflessly work long hours to make sure the mission is accomplished. I also learned early on that the air force is like family. I have always felt taken care of no matter where I have been in the world so long as my military family was close by.

Q. Please tell us briefly about a typical day in your life in the military.

A. There is no "typical" day in the military. Every assignment I have had has involved a wide variety of interesting and challenging issues. I started as a prosecutor, which is referred to as a chief of military justice in the air force. Then I switched to being an *Area Defense Counsel,* which is like a public defender in the civilian sector. Next I went on to teach in the law department at the United States Air Force Academy, which I did for three years and absolutely loved.

That assignment was followed by one of the most challenging jobs when I served as the legal adviser to the superintendent and commandant at the United States Air Force Academy during a time when the academy was being scrutinized for the practices, policies, and procedures in place for responding to reports of sexual assault.

That two-year assignment was followed by the most rewarding position in my active duty career. I was assigned to serve as the deputy staff judge advocate at Osan Air Base, Republic of Korea. This was a tough assignment, but very rewarding because most people at this assignment were away from family. We practiced how to respond in the event North Korea ever attacked us, and that training and practice was intense. We had to be ready to get into our full chemical gear within minutes and learn many new things, while still balancing and completing the workload in one of the busiest offices in the air force.

Now, I am an air force reservist. Once a month I travel to my base of assignment and work throughout the weekend. We provide legal assistance to fellow Reserve unit members (wills, powers of attorney, advice on divorce, child custody, etc.), and we also advise commanders and supervisors on how to legally and appropriately address illegal behavior or disciplinary problems. A typical day most anywhere I have been involves research of the air force regulations (of which there are a few hundred), the law and applying the law, rules, or regulations to the specific fact pattern. Air force attorneys spend much of their time advising commanders on the rule of law and how to appropriately apply it in a wide variety of situations.

Q. What is one thing that young people may not know about a career in the military?

A. The military teaches you how to lead, influence, and inspire those around you. If you apply yourself, align yourself with positive mentors, and regularly schedule time for self-reflection and initiate feedback sessions with your mentors you will learn to be a top leader, manager, and supervisor. I have found that my skills are remarkably more advanced than my civilian counterparts, and I know it is the reason I was able to find a civilian job as a senior, supervisory attorney after serving 10 years on active duty in the air force.

Q. What are the most important personal and professional qualities for military lawyers?

A. Integrity; selflessness; and a willingness to help others more than you look for opportunities to help or advance yourself.

Q. What advice would you give to young people who are interested in the field?

A. If you are interested in a career in the military, join an ROTC program while in college and see if you think you are a good fit. I did that and learned very early on that I wanted to be a part of the air force. Choosing my career path as an attorney came later, and I have never regretted that decision. I have become a confident person who knows how to get the best out of the people who work for me.

I would also advise that getting to my level took a lot of hard work, attention to detail, and an ongoing commitment to self-reflection. I have filled book upon journal book with my lessons learned. I have also spent a lot of time planning how I would handle a tough conversation with a subordinate/employee in my journals. This has paid off, and it has taught me how very important it is to reflect on what you will say and how you will deliver important information to your employees. I believe by doing this I have been rewarded in many circumstances by turning people around (motivating people to change for the better; to better themselves in ways that they did not even think possible). Even when I had to fire someone I was able to maintain a positive relationship with that individual once he was reassigned. In my opinion, those things can't happen unless you spend the time identifying and clearly articulating expectations, reflecting on and deciding on the right way to deliver that message, and following up as needed.

Mechanic and Repair Technologists and Technicians

QUICK FACTS

School Subjects
Computer science
Technical/shop

Personal Skills
Mechanical/manipulative
Technical/scientific

Work Environment
Indoors and outdoors
Primarily multiple locations

Minimum Education Level
Varies by career specialty

Salary Range
$15,617 to $30,618 to $66,154 (enlisted personnel)
$25,358 to $70,877 to $174,103 (officers)

Outlook
About as fast as the average

DOT
003, 600, 601, 620, 621, 625, 637, 710, 806, 810, 899, 952

GOE
05.01.01, 05.02.01, 05.02.02, 05.03.01, 05.03.02, 05.03.03, 08.04.01, 08.06.01

(continues)

OVERVIEW

The military relies on *mechanic and repair technologists and technicians* to keep tanks, trucks, artillery, naval guns systems, infantry weapons, airplanes, helicopters, submarines, ships, boilers and furnaces, generators, medical laboratory equipment, fuel systems, gas and diesel engines, and other equipment working properly. Workers in this field have excellent mechanical aptitude and are proficient in the use of hand and power tools. Opportunities are available for both enlisted personnel and officers in the U.S. Air Force, Army, Coast Guard, Marines, and Navy. There were more than 173,000 mechanic and repair technologists and technicians employed in the military in 2005.

HISTORY

Throughout history, the strength of a nation's armed forces can be directly linked to the quality of its repair and service personnel. During the Revolutionary War, for example, mechanics and repairers were needed to keep the supply carts drawn by horses working properly, as well as maintain guns, cannons, and meteorological and surveying instruments. As technology advanced, U.S. military mechanics and repairers faced larger challenges such as maintaining aircraft, jeeps, radios, health care equipment, complicated electronics, and entire power plants in order to ensure that soldiers were successful on the battlefield and beyond. Today, military mechanic and repair technologists and technicians repair and maintain thousands of types of equipment and technologies—ranging from EEG monitors, periscopes, and power saws; to jets, ships, and submarines; to entire nuclear power plants.

QUICK FACTS

(continued)

O*NET-SOC

17-3023.02, 17-3024.00, 17-3027.00, 47-2111.00, 49-2091.00, 49-3011.00, 49-3011.01, 49-3011.02, 49-3011.03, 49-3023.00, 49-3031.00, 49-3051.00, 49-9021.00, 49-9021.01, 49-9021.02, 49-9042.00, 49-9043.00, 49-9062.00, 49-9069.99, 49-9092.00, 51-4041.00, 51-4061.00, 51-4072.00, 51-4111.00, 51-4121.00, 51-4121.01, 51-4121.02, 51-4121.03, 51-4122.00, 51-4122.01, 51-4122.02, 51-8011.00, 51-8012.00, 51-8013.00, 51-8013.01, 51-8013.02

THE JOB

Many opportunities exist for mechanics and related workers in the military. The following paragraphs describe some of the more popular career options.

Aircraft mechanics examine, service, repair, and overhaul aircraft and helicopters and their engines. They also repair, replace, and assemble parts of the airframe (the structural parts of the plane other than the power plant or engine).

Automotive and heavy equipment mechanics maintain and repair cars, trucks, jeeps, tanks and other combat vehicles, and construction equipment (such as bulldozers and power shovels). Using both hand tools and specialized diagnostic test equipment, they pinpoint problems and make the necessary repairs or adjustments. In addition to performing complex and difficult repairs, technicians perform a number of routine maintenance procedures, such as oil changes, tire rotation, and battery replacement.

Divers perform a wide variety of underwater jobs, both in oceans and in fresh water, using special underwater breathing equipment. They perform many tasks including search and rescue operations, constructing piers and other structures, ship repair and service, and patrolling shipyards and ports to ensure that terrorists do not try to sabotage ships, equipment, or facilities. *Scuba divers* are specialized divers who work just below the surface. *Deep sea divers* work at depths of up to 300 feet below the surface.

Electrical products repairers maintain and repair electric tools (such as saws and drills), electric motors, and medical equipment and instruments (such as heart-lung machines, artificial kidney machines, dental equipment, patient monitors, chemical analyzers, and other electrical, electronic, mechanical, or pneumatic devices). They disassemble equipment to locate malfunctioning components, repair or replace defective parts, and reassemble the equipment, adjusting and calibrating it to ensure that it operates according to manufacturers' specifications.

Heating and cooling mechanics work on systems that control the temperature, humidity, and air quality of enclosed environments in military buildings, ships, and airplanes. They help design, manufacture, install, and maintain climate-control equipment.

Machinists use machine tools, such as drill presses, lathes, and milling machines, to repair or produce metal parts that meet precise specifications. They combine their knowledge of metals with skillful handling of machine tools to make precision-machined products.

Marine engine mechanics inspect, maintain, and repair marine vessels, from small boats to military destroyers. They work on vessel hulls, gasoline and diesel engines, transmissions, navigational equipment, and electrical, propulsion, and refrigeration systems.

Nondestructive testers examine the metal parts of military vehicles and equipment for any damages or weaknesses caused by prolonged use or through battle. They may check for cracks in the skin of an air force jet, or the weakening on the hull of a navy frigate ship. Using methods such as radiography, infrared thermograph, or liquid penetrants, testers are able to detect many potentially dangerous fatigue-induced weaknesses without actually damaging the test subject. Their duties include conducting inspections, operating testing equipment, and preparing reports of their findings.

Power plant electricians maintain and operate equipment that generates the electricity needed to power all military bases. They work with the stationary power plant equipment of permanent military bases, as well as the smaller mobile units used to power military field installations. Their duties include the maintenance and repair of motors, switchboards, and control equipment. Electricians also make routine inspections and repair power circuits and electrical fixtures, as well as locate ground wires and short circuits.

Power plant operators monitor and operate the machinery (nuclear reactors, oilers, turbines, and portable generators) that generates electric power for use on ships, submarines, and military bases. They also operate switches that control the amount of power created by the various generators and regulate the flow of power to outgoing transmission lines. They keep track of power demands on the system and respond to changes in demand by turning generators on and off and connecting and disconnecting circuits.

Military bases, ships, and field installations receive electricity from power-generating stations, also known as powerhouses. *Powerhouse*

Rank and Military Branch by Occupation

Job Title	Rank	Military Branches
Aircraft Mechanics	Enlisted	Air Force, Army, Coast Guard, Marines, Navy
Automotive and Heavy Equipment Mechanics	Enlisted	Air Force, Army, Coast Guard, Marines, Navy
Divers	Enlisted	Army, Coast Guard, Marines, Navy
Electrical Products Repairers	Enlisted	Air Force, Army, Coast Guard, Marines, Navy
Heating and Cooling Mechanics	Enlisted	Air Force, Army, Coast Guard, Marines, Navy
Machinists	Enlisted	Air Force, Army, Coast Guard, Marines, Navy
Marine Engine Mechanics	Enlisted	Air Force, Army, Coast Guard, Marines, Navy
Nondestructive Testers	Enlisted	Air Force, Coast Guard, Marines, Navy
Power Plant Electricians	Enlisted	Air Force, Army, Coast Guard, Marines, Navy
Power Plant Operators	Enlisted	Army, Coast Guard, Marines, Navy
Powerhouse Mechanics	Enlisted	Air Force, Army, Coast Guard, Marines, Navy
Precision Instrument and Equipment Repairers	Enlisted	Coast Guard, Navy
Ship Electricians	Enlisted	Coast Guard, Navy
Survival Equipment Specialists	Enlisted	Air Force, Army, Coast Guard, Marines, Navy
Transportation Maintenance Managers	Officers	Air Force, Army, Coast Guard, Marines, Navy
Weapons Maintenance Technicians	Enlisted	Air Force, Army, Coast Guard, Marines, Navy
Welders and Metal Workers	Enlisted	Air Force, Army, Coast Guard, Marines, Navy

Source: U.S. Department of Defense

mechanics install, maintain, and repair electrical and mechanical equipment used in power-generating stations. They specialize according to the source of power, ranging from small generators run by gasoline to massive nuclear reactors. Their duties include installing and maintaining equipment such as engines, pumps, turbines, and air compressors. They use many kinds of tools to do their work, including timing lights, pressure gauges, and small hand tools.

Precision instrument and equipment repairers maintain, adjust, and repair a variety of instruments used by the military, including photographic and imaging equipment such as cameras, film processing equipment, and projectors; instruments that measure altitude, temperature, distance, pressure, underwater depth, and other physical properties; watches, timers, and clocks; and weapon-aiming devices such as range finders, periscopes, and telescopes.

Ship electricians specialize in the maintenance and repair of all power-generating equipment aboard military ships and submarines. They maintain circuits, transformers, regulators, and other devices that supply electricity throughout a ship. They also mount wiring for light and other equipment. In addition, their duties include troubleshooting for potential problems with a ship's power plant, electrical wiring, and electrical devices for a nuclear reactor.

During times of emergency, military personnel rely on the use of survival equipment to save their lives. *Survival equipment specialists* maintain and repair survival equipment such as parachutes, water rafts, and life support equipment. For example, specialists inspect parachutes for rips or tangled lines, and make necessary repairs before properly packing them. Their duties include training military personnel in the proper use of the equipment. They are also in charge of stocking life rafts and aircrafts with emergency provisions such as flare guns, fire extinguishers, and first-aid kits.

Transportation maintenance managers oversee all maintenance done for the military's transportation equipment. They direct mechanics and technicians specialized in the repair and maintenance of ships, aircrafts, trucks, and other land vehicles. A manager's duties include maintaining an inventory of tools and spare parts, keeping maintenance records for work done, and training and scheduling staff. They may also develop standards and policies for work done in their department.

The military uses many different weapons during combat and training exercises. *Weapons maintenance technicians* are responsible for the repair and upkeep of weapons ranging from field weapons such as machine guns to long-range nuclear warheads launched from a submarine. Their duties include the testing and repair of support systems such as launchers, fixed or mobile mounts, or tripods. Technicians receive specialized training according to the type of weapon they service.

Welders and metal workers operate a variety of special equipment to join metal parts together permanently, usually using heat and sometimes pressure. They construct and repair ships, submarines, tanks, aircraft, jeeps, automobiles, buildings, bridges, highways, and many other metal structures and manufactured products.

REQUIREMENTS

High School

High school courses that will prepare you for these occupations include mechanical drawing, metal shop, electrical shop, woodworking, blueprint reading, general science, computer science, and applied mathematics.

Postsecondary Training

Educational requirements in the civilian sector vary by career, but most workers receive their training by participating in an apprenticeship or attending a technical college. Managers typically earn a bachelor's degree in business, management, or their particular specialty such as electronics, aviation, or nuclear engineering.

In the military, enlisted personnel (which include every occupation listed in this article except transportation maintenance managers) train via classroom instruction, on-the-job experience, and advanced course work. Course work varies by specialty. For example, aircraft mechanics take classes in engine repair and disassembly and repair of aircraft systems, airframes, and coverings. They also perform hands-on inspection and repair of aircraft engines. Survival equipment specialists take classes that teach them how to prepare parachutes for use, maintain oxygen equipment, and repair inflatable rafts and other survival equipment. Divers take classes in the principles of scuba diving, equipment maintenance, and welding and cutting underwater. They also constantly practice their diving and repair skills in order to master their craft and be ready when called on to perform dangerous tasks.

As officers, transportation maintenance managers must enter the military with at least a bachelor's degree. Once they join the military, they take classes to learn how to manage aircraft or aircraft electronics maintenance; oversee the maintenance of vehicle, railroad, and other equipment; and use management information systems.

Visit the U.S. Department of Defense's Web site, http://www.todays military.com, for more on military training for those interested in becoming mechanic and repair technologists and technicians.

Other Requirements

To be successful in this field, workers must have the ability to use hand and power tools, be attentive to detail, have an interest in troubleshooting and solving mechanical and other types of problems, and enjoy working with their hands. Transportation maintenance managers also need to have these skills, as well as enjoy directing and planning the work of others. They should be decisive and have good leadership abilities. Nondestructive testers should have a keen eye for detail, de dependable, and be able to operate testing equipment. Divers should be self-reliant, decisive, and be in excellent physical shape.

Visit the U.S. Department of Defense's Web site, http://www.todaysmilitary.com, for more on personal requirements for workers in these careers.

EXPLORING

Working with electronic kits, tinkering with automobile engines, and assembling model airplanes are good ways of gauging your ability to do the kinds of work performed by mechanic and repair technologists and technicians. A guided tour of an airfield, automotive or truck repair shop, marine engine repair facility, power plant, or any other place that mechanics work can give you a good introduction to the field. Using hand and power tools to work on household projects is another excellent way to get experience in the field.

Others ways to learn about any of the jobs listed in this article include reading books and magazines about these careers, visiting Web sites of professional associations, and asking your guidance

Two mechanics inspect an F-15C Eagle aircraft at Kadena Air Base, Japan. *(Airman 1st Class Kelly Timney, U.S. Air Force, U.S. Department of Defense)*

counselor or teacher to arrange an information interview with a worker in the field.

To learn more about career opportunities in the military, visit the Web sites listed at the end of this article.

EMPLOYERS

The U.S. government employs the military. In 2005 more than 173,000 mechanic and repair technologists and technicians were employed in the military.

STARTING OUT

Contact a military recruiter for more information about the wide variety of careers available in the armed forces. Visit the Web sites listed at the end of this article to locate a recruiting office near you. To start out in any branch, you will need to pass physical and medical tests, the Armed Services Vocational Aptitude Battery exam, and basic training.

ADVANCEMENT

Each military branch has nine enlisted grades (E-1 through E-9) and 10 officer grades (O-1 through O-10). The higher the number is, the more advanced a person's rank is. The various branches of the military have somewhat different criteria for promoting individuals; in general, however, promotions depend on factors such as length of time served, demonstrated abilities, recommendations, and scores on written exams. Promotions become more and more competitive as people advance in rank. On average, a diligent enlisted person can expect to earn one of the middle noncommissioned or petty officer rankings (E-4 through E-6); some officers can expect to reach lieutenant colonel or commander (O-5). Outstanding individuals may be able to advance beyond these levels.

EARNINGS

The U.S. Congress sets the pay scales for the military after hearing recommendations from the president. The pay for equivalent grades is the same in all services (that is, anyone with a grade of E-4, for example, will have the same basic pay whether in the army, navy, marines, air force, or Coast Guard). In addition to basic pay, personnel who frequently and regularly participate in combat may earn hazardous duty pay. Other special allowances include special duty pay and foreign duty pay. Earnings start relatively low but increase on a fairly regular basis as individuals advance in rank. See the appendix at the end of this book for detailed information on pay scales for the U.S. military. When reviewing earnings, it is important to keep in mind that members of the military receive free housing, food, and health care—items that civilians typically pay for themselves.

Additional benefits for military personnel include uniform allowances, 30 days' paid vacation time per year, and the opportunity to retire after 20 years of service. Generally, those retiring will receive 40 percent of the average of the highest three years of their base pay. This amount rises incrementally, reaching 75 percent of the average of the highest three years of base pay after 30 years of service. All retirement provisions are subject to change, however, and you should verify them as well as current salary information before you enlist. Those who retire after 20 years of service are usually in their 40s and thus have plenty of time, as well as an accumulation of skills, with which to start a second career.

WORK ENVIRONMENT

The work environment for mechanic and repair technologists and technicians depends a great deal on their specialty. For example, aircraft mechanics work in aircraft hangars and machine shops on air bases or aboard aircraft carriers. When doing overhauling and major inspection work, they generally work in hangars with adequate heat, ventilation, and lights. If the hangars are full, however, or if repairs must be made quickly, they may work outdoors, sometimes in unpleasant weather. Divers work underwater and on ships planning and preparing for their assignments. Ship electricians typically work indoors aboard submarines or ships, but also work in shops on land. Working conditions for technologists and technicians can sometimes be noisy and dirty. The work is sometimes physically strenuous and demanding. Mechanics often have to lift or pull as much as 70 pounds of weight. They may stand, lie, or kneel in awkward positions, sometimes in precarious places such as on a scaffold or ladder.

OUTLOOK

Employment in the armed forces is expected to grow about as fast as the average for all occupations through 2014, according to the U.S. Department of Labor. When the economy is good and/or during times of war, more people pursue employment in the civilian workforce, which creates additional opportunities in the military. With the U.S. military involved in several international conflicts, most significantly in Iraq and Afghanistan, demand should continue to be strong for military workers.

Opportunities should be good for mechanic and repair technologists and technicians. Ongoing military operations in Iraq and

Afghanistan have created considerable wear and tear on military equipment, and mechanics will be needed to keep this equipment in operational condition.

The employment outlook in the civilian sector varies by occupation. The U.S. Department of Labor offers the following predictions for professions in the field: aircraft mechanics and automotive, heavy equipment mechanics, marine engine mechanics, power plant electricians, precision instrument and equipment repairers, and ship electricians, about as fast as the average; heating and cooling mechanics, faster than the average; electrical products repairers and machinists and welders and metal workers, more slowly than the average; and power plant operators, decline.

FOR MORE INFORMATION

For general information on the power industry, contact

American Public Power Association
1875 Connecticut Avenue, NW, Suite 1200
Washington, DC 20009-5715
Tel: 202-467-2900
http://www.appanet.org

For more information about becoming a welder, contact

American Welding Society
550 NW LeJeune Road
Miami, FL 33126-5649
Tel: 800-443-9353
http://www.aws.org

For information on diesel education, contact

Association of Diesel Specialists
10 Laboratory Drive
PO Box 13966
Research Triangle Park, NC 27709-3966
Tel: 919-406-8804
Email: info@diesel.org
http://www.diesel.org

For information about union membership for electricians, contact

International Brotherhood of Electrical Workers
900 Seventh Street, NW
Washington, DC 20001-3886
Tel: 202-833-7000
http://www.ibew.org

For information on careers, training, and student membership, contact

ISA—The Instrumentation, Systems, and Automation Society
67 Alexander Drive
Research Triangle Park, NC 27709
Tel: 919-549-8411
Email: info@isa.org
http://www.isa.org

For information on certified educational programs and careers, contact

National Automotive Technicians Education Foundation
101 Blue Seal Drive, Suite 101
Leesburg, VA 20175-5646
Tel: 703-669-6650
http://www.natef.org

For industry information, contact

National Marine Electronics Association
7 Riggs Avenue
Severna Park, MD 21146-3819
Tel: 410-975-9425
Email: info@nmea.org
http://www.nmea.org

For information about training and opportunities in the precision machining and metalworking industries, contact the following organizations:

National Tooling and Machining Association
9300 Livingston Road
Fort Washington, MD 20744-4988
Tel: 800-248-6862
http://www.ntma.org

Precision Machined Products Association
6700 West Snowville Road
Brecksville, OH 44141-3212
Tel: 440-526-0300
http://www.pmpa.org

For information on diving, contact

PADI
30151 Tomas Street
Rancho Santa Margarita, CA 92688-2125
Tel: 800-729-7234
http://www.padi.com

For information on union membership for plumbers and related workers, contact

Plumbing-Heating-Cooling Contractors Association
180 South Washington Street
PO Box 6808
Falls Church, VA 22046-2900
Tel: 800-533-7694
Email: naphcc@naphcc.org
http://www.phccweb.org

For information on aviation maintenance and scholarships, contact

Professional Aviation Maintenance Association
400 Commonwealth Drive
Warrendale, PA 15096-0001
Tel: 866-865-7262
Email: hq@pama.org
http://www.pama.org

To get information on specific branches of the military, check out this site, which is the home of ArmyTimes.com, NavyTimes.com, AirForceTimes.com, and MarineCorpsTimes.com:

Military City
http://www.militarytimes.com

If you're thinking of joining the armed forces, take a look at this site, which guides students and parents through the decision-making process:

Today's Military
http://www.todaysmilitary.com

For information on military careers, contact

United States Air Force
http://www.airforce.com

United States Army
http://www.goarmy.com

United States Coast Guard
http://www.gocoastguard.com

United States Marine Corps
http://www.marines.com

United States Navy
http://www.navy.com

Media and Public Affairs Occupations

QUICK FACTS

School Subjects
English
Journalism
Speech

Personal Skills
Communication/ideas
Helping/teaching

Work Environment
Indoors and outdoors
Primarily multiple locations
Varies by career specialty

Salary Range
$15,617 to $30,618 to $66,154 (enlisted personnel)
$25,358 to $70,877 to $174,103 (officers)

Outlook
About as fast as the average

DOT
131, 165, 184, 194

GOE
01.01.01, 01.02.01, 01.03.01, 01.05.01, 01.08.01, 13.01.01

O*NET-SOC
27-2012.02, 27-2012.03, 27-3020.00, 27-3021.00, 27-3022.00, 27-4011.00, 27-4012.00, 27-4014.00

OVERVIEW

Media and public affairs workers in the military provide information about the military via newspapers, broadcasts, Web sites, and press conferences. Opportunities for both enlisted personnel and officers are available in the U.S. Air Force, Army, Coast Guard, Marines, and Navy.

HISTORY

The first American newspaper, published in 1690, was suppressed four days after it was published. And it was not until 1704 that the first continuous newspaper appeared. A number of developments in the printing industry made it possible for newspapers to be printed more cheaply. In the late-19th century, new types of presses were developed to increase production, and more importantly, the Linotype machine was invented. The Linotype mechanically set letters so that handset type was no longer necessary. This dramatically decreased the amount of prepress time needed to get a page into print. Newspapers could respond to breaking stories more quickly, and late editions with breaking stories became part of the news world. These technological advances, along with an increasing population, factored into the rapid growth of the newspaper industry in the United States. Only 37 newspapers existed in the United States in 1776. Today there are more than 1,450 daily and more than 6,700 weekly newspapers in the country. As newspapers grew in size and widened the scope of their coverage, it became necessary to increase the number of employees and to assign them specialized jobs. Reporters have always been the heart of newspaper staffs. However, in today's complex world, with the public hungry for news as it occurs, reporters, newswriters, and correspondents are involved in all media—not only newspapers, but magazines, radio, and television as well. Today, with the advent of the Internet, many newspapers are going online, causing many reporters to become active participants on the Information Superhighway.

Broadcast technology developed in the 20th century and enabled people to reach large audiences around the world instantly, changing forever the way we communicate. In the early 1900s, transmitting and receiving devices were relatively simple, and hundreds of amateurs constructed transmitters and receivers on their own and experimented with radio. Ships were rapidly equipped with radios so they could communicate with each other and with shore bases while at sea. In 1906, the human voice was transmitted for the first time. Small radio shows started in 1910. Ten years later, two commercial radio stations went on the air, and by 1921, a dozen local stations were broadcasting. Four years later, the first radio broadcast was transmitted around the world. Although the advent of television changed the kind of programming available on the radio (from comedy, drama, and news programs to radio's current schedule of music, phone-in talk shows, and news updates), there has been a steady growth in the number of radio stations in the United States. According to the *CIA World Factbook,* 13,750 radio stations were broadcasting in the United States in 2006.

Modern television developed from experiments with electricity and vacuum tubes in the mid-1800s, but it was not until 1939, when President Franklin Roosevelt used television to open the New York World's Fair, that the public realized the power of television as a means of communication. Several television stations went on the air shortly after this demonstration and successfully televised professional baseball games, college football games, and the Republican and Democratic conventions of 1940. The onset of World War II limited the further development of television until the war was over. Since television's strength is the immediacy with which it can present information, news programs became the foundation of regular programming. *Meet the Press* premiered in 1947, followed by nightly newscasts in 1948. The industry expanded rapidly in the 1950s. The Federal Communications Commission lifted a freeze on the processing of station applications, and the number of commercial stations grew steadily, from 120 in 1953 to more than 2,200 stations in 2006, according to the *CIA World Factbook.*

The U.S. military has used radio, television, and print mediums to provide information to the public and military personnel for years, as well as to try to reach people living in foreign countries under dictatorships or repressive governments. Military-oriented newspapers were available in the early days of our nation, and continue to provide military personnel with news and entertainment today. (For example, *Stars and Stripes* has been published continuously for military personnel in Europe since 1942, and since 1945 for those in the Pacific.) The Armed Forces Radio Service was founded by the War Department in 1942. A service station in the Panama Canal Zone provided the first broadcast. A television service was introduced in 1954, with the organization changing its name to the Armed Forces Radio and Television Service. In 1998, the organization's name was changed again to the American Forces Network. Today, the network offers a variety of radio and television programming to military personnel throughout the world.

THE JOB

Some of the more popular options for those interested in media and public affairs careers in the military are described below.

Audiovisual and broadcast directors plan and organize entertainment programs, news broadcasts, training films, and other material that the military produces. They may be directly involved with filming or broadcasting—such as preparing scripts, determining camera angles, or directing the actors and crew during a production—or they may manage other directors.

Audiovisual and broadcast technicians assist in the production of training, entertainment, and news programs, and with the recording of military operations, ceremonies, and other events. They operate and maintain the electronic equipment used to record and transmit the audio for radio signals and the audio and visual images for television signals, and may also help to prepare scripts, design or arrange background scenery, and plan special effects. Audiovisual and broadcast technicians may work in a broadcasting station or at an outside site as a field technician.

Broadcast journalists and newswriters are the foot soldiers for the military's newspapers, magazines, and television and radio broadcast operations. They gather and analyze information about current events, conduct interviews, and write stories for publication or scripts for broadcasting. They collect information on newsworthy people and events in the military and prepare stories for newspaper or magazine publication or for radio or television broadcast. They may also be involved with choosing photographs, maps, or other illustrations that may enhance an article or broadcast.

Public information officers provide information to the public, usually via media outlets, about military events, goals, and accomplishments, with the goal of presenting a positive image of the military. They may develop and maintain programs that inform the

Rank and Military Branch by Occupation

Job Title	Rank	Military Branches
Audiovisual and Broadcast Directors	Officer	Air Force, Army, Marines, Navy
Audiovisual and Broadcast Technicians	Enlisted	Air Force, Army, Coast Guard, Marines, Navy
Broadcast Journalists and Newswriters	Enlisted	Air Force, Army, Coast Guard, Marines, Navy
Public Information Officers	Officer	Air Force, Army, Coast Guard, Marines, Navy

Source: U.S. Department of Defense

public about military activities, prepare news releases, answer the questions of government employees or private citizens, arrange for interviews between members of the media and military personnel, and supervise other public information workers.

REQUIREMENTS

High School

High school courses that will provide you with a firm foundation for a media or public affairs career include English, journalism, history, social studies, communications, typing, and computer science. Speech courses will help you hone your interviewing skills, which are necessary for success as a reporter. In addition, it will be helpful to take college prep courses, such as foreign language, math, and science.

Postsecondary Training

Educational requirements in the civilian sector vary by career. For example, audiovisual and broadcast directors typically have a bachelor's degree in radio and television production and broadcasting, communications, liberal arts, or business administration. Audiovisual and broadcast technicians have an associate's or bachelor's degree in broadcast engineering. Journalists and newswriters have at least a bachelor's degrees in journalism. Public information officers typically have a bachelor's degree in public relations, English, or journalism.

Military workers in this specialty receive job training in classroom settings, on the job, and through advanced courses. Course work varies by career. For example, audiovisual and broadcast technicians learn how to operate motion picture and audio equipment, create scripting and special effects techniques, and maintain public address sound equipment. Broadcast journalists and newswriters take classes that teach them how to write clearly and concisely for print or broadcast, research stories, and conduct interviews.

Visit the U.S. Department of Defense's Web site, http://www.todaysmilitary.com, for more on military training for those interested in media and public affairs careers.

Other Requirements

Personal requirements for workers in these occupations vary by specialty. For example, audiovisual and broadcast workers must be very organized, be comfortable working as a member of a team, and be adept at working with film and audio equipment. Broadcast journalists and writers must be very attentive to detail, have excellent writing skills, and be able to gather information quickly for publication or broadcast. Public information officers should be very good at public speaking, have excellent writing skills, and be interested in current events.

Visit the U.S. Department of Defense's Web site, http://www.todaysmilitary.com, for more on personal requirements for workers in these careers.

EXPLORING

You can explore a career as an audiovisual and broadcast director, technician, or journalist in a number of ways. You can talk to professionals at local newspapers and radio and TV stations. You can interview the admissions counselor at the school of journalism closest to your home. High school students can acquire practical experience by working on school newspapers or at school radio or television stations.

Aspiring public information officers can write for their school newspaper or volunteer to do public relations work for a local nonprofit organization.

Others ways to learn about any of the jobs listed in this article include reading books and magazines about these careers (see the sidebar, "Magazines of the Armed Forces," for a list of military magazines), visiting Web sites of professional associations, arranging a tour of newsroom or broadcast studio, and asking your guidance counselor or teacher to arrange an information interview with a reporter, broadcast professional, or public information officer.

To learn more about career opportunities in the military, visit the Web sites listed at the end of this article.

EMPLOYERS

The U.S. government employs the military. No specific employment statistics are available for military media and public affairs workers.

Overall, 1.4 million men and women are on active duty and another 1.2 million volunteers serve in the Guard and Reserve.

STARTING OUT

Contact a military recruiter to learn more about this career. Visit the Web sites listed at the end of this article to locate a recruiting office near you. To start out in any branch, you will need to pass physical and medical tests, the Armed Services Vocational Aptitude Battery exam, and basic training.

ADVANCEMENT

Each military branch has nine enlisted grades (E-1 through E-9) and 10 officer grades (O-1 through O-10). The higher the number is, the more advanced a person's rank is. The various branches of the military have somewhat different criteria for promoting individuals; in general, however, promotions depend on factors such as length of time served, demonstrated abilities, recommendations, and scores on written exams. Promotions become more and more competitive as people advance in rank. On average, a diligent enlisted person can expect to earn one of the middle noncommissioned or petty officer rankings (E-4 through E-6); some officers can expect to reach lieutenant colonel or commander (O-5). Outstanding individuals may be able to advance beyond these levels.

EARNINGS

The U.S. Congress sets the pay scales for the military after hearing recommendations from the president. The pay for equivalent grades is the same in all services (that is, anyone with a grade of E-4, for example, will have the same basic pay whether in the army, navy, marines, air force, or Coast Guard). In addition to basic pay, personnel who frequently and regularly participate in combat may earn hazardous duty pay. Other special allowances include special duty pay and foreign duty pay. Earnings start relatively low but increase on a fairly regular basis as individuals advance in rank. See the appendix at the end of this book for detailed information on pay scales for the U.S. military. When reviewing earnings, it is important to keep in mind that members of the military receive free housing, food, and health care—items that civilians typically pay for themselves.

Additional benefits for military personnel include uniform allowances, 30 days' paid vacation time per year, and the opportunity to retire after 20 years of service. Generally, those retiring will receive 40 percent of the average of the highest three years of their base pay.

Magazines of the Armed Forces

All Services
Stars and Stripes
http://www.estripes.com

Air Force
Air & Space Power
http://www.airpower.maxwell.af.mil/airchronicles/apje.html

Airman Quarterly
http://www.af.mil/news/airman

Citizen Airman
http://www.citamn.afrc.af.mil

Army
Army Engineer
http://www.armyengineer.com/AEA_Magazine.htm

Military Review
http://usacac.army.mil/CAC/milreview/English/english.asp

Parameters
http://www.carlisle.army.mil/usawc/Parameters

Soldiers
http://www.army.mil/publications

Coast Guard
Coast Guard
http://www.uscg.mil/magazine

Coast Guard Reservist
http://www.uscg.mil/hq/reserve/magazine/magazine.htm

Marines
Continental Marine Magazine
http://www.marforres.usmc.mil/News

Marines
http://www.mcnews.info/mcnewsinfo/marines/gouge

Navy
All Hands
http://www.navy.mil/allhands.asp?x=search

Naval Aviation News
http://www.history.navy.mil/branches/nhcorg5.htm

The Navy Reservist
http://navyreserve.navy.mil

Seabee
https://www.seabee.navy.mil

This amount rises incrementally, reaching 75 percent of the average of the highest three years of base pay after 30 years of service. All retirement provisions are subject to change, however, and you should verify them as well as current salary information before you enlist. Those who retire after 20 years of service are usually in their 40s and thus have plenty of time, as well as an accumulation of skills, with which to start a second career.

WORK ENVIRONMENT

Those in media and public affairs occupations typically work in offices or studios. Audiovisual and broadcast directors and technicians and broadcast journalists sometimes work outdoors in extreme weather conditions. Audiovisual and broadcast directors may direct film crews in combat zones.

OUTLOOK

Employment in the armed forces is expected to grow about as fast as the average for all occupations through 2014, according to the U.S. Department of Labor. When the economy is good and/or during times of war, more people pursue employment in the civilian workforce, which creates additional opportunities in the military. With the U.S. military involved in several international conflicts, most significantly in Iraq and Afghanistan, demand should continue to be strong for military workers.

Opportunities should be good for media and public affairs workers in the military. As a result of bad publicity and scandals, such as the mistreatment of civilians in war zones and military detainees in prisons, the military is increasingly concerned with presenting a positive image to the public. The military will also continue to entertain and enlighten its personnel via the American Forces Network and other methods.

The employment outlook in the civilian sector varies by occupation. The U.S. Department of Labor offers the following predictions for professions in the field: audiovisual and broadcast technicians, about as fast as the average; public information officers, faster than the average; and audiovisual and broadcast directors and broadcast journalists and newswriters, more slowly than the average.

FOR MORE INFORMATION

This organization provides general educational information on all areas of journalism, including newspapers, magazines, television, and radio.

Association for Education in Journalism and Mass Communication
234 Outlet Pointe Boulevard
Columbia, SC 29210-5667
Tel: 803-798-0271
Email: aejmchq@aejmc.org
http://www.aejmc.org

For a list of schools offering degrees in broadcasting, contact
Broadcast Education Association
1771 N Street, NW
Washington, DC 20036-2891
Tel: 888-380-7222
Email: beainfo@beaweb.org
http://www.beaweb.org

To get information on specific branches of the military, check out this site, which is the home of ArmyTimes.com, NavyTimes.com, AirForceTimes.com, and MarineCorpsTimes.com:
Military City
http://www.militarytimes.com

If you're thinking of joining the armed forces, take a look at this site, which guides students and parents through the decision-making process:
Today's Military
http://www.todaysmilitary.com

For information on military careers, contact
United States Air Force
http://www.airforce.com

United States Army
http://www.goarmy.com

United States Coast Guard
http://www.gocoastguard.com

United States Marine Corps
http://www.marines.com

United States Navy
http://www.navy.com

For information on military broadcasting and journalism careers, visit the following Web site:
American Forces Information Service
http://www.defenselink.mil/home/news_products.html

Military Pilots

OVERVIEW

Military pilots fly various types of specialized aircraft to transport troops and equipment and to execute combat missions. Military aircraft make up of one of the world's largest fleets of specialized airplanes.

Pilots within the five branches of the U.S. armed forces train, organize, and equip the nation's air services to support the national and international policies of the government. There are approximately 16,000 airplane pilots and 6,500 helicopter pilots in the military.

QUICK FACTS

School Subjects
Computer science
Government
Physics

Personal Skills
Leadership/management
Technical/scientific

Work Environment
Indoors and outdoors
Primarily multiple locations

Minimum Education Level
Bachelor's degree

Salary Range
$25,358 to $70,877 to $174,103 (officers)

Outlook
About as fast as the average

DOT
378, 632

GOE
04.05.01

O*NET-SOC
55-1011.00, 55-2011.00, 55-3011.00, 55-3017.00

HISTORY

The age of modern aviation is generally considered to have begun with the famous flight of Orville and Wilbur Wright's heavier-than-air machine on December 17, 1903. On that day, the Wright brothers flew their machine four times and became the first airplane pilots.

Aviation developed rapidly as designers raced to improve upon the Wright brothers' design. During the early years of flight, many aviators earned a living as "barnstormers," entertaining people with stunts and by taking passengers on short flights around the countryside. Airplanes were quickly adapted to military use. In 1907, the army created an aeronautical division. Air power proved invaluable a few years later during World War I, bringing about major changes in military strategy. As a result, the United States began to assert itself as an international military power, and accordingly, the Army Air Service was created as an independent unit in 1918, although it remained under army direction for a time.

With the surprise attack on Pearl Harbor in 1941, America was plunged into World War II. At its height, 13 million Americans fought in the different branches of the military services. When the war ended, the United States emerged as the strongest military power in the Western world. A large part of America's military success was due to the superiority of its air forces. Recognition of the strategic importance of air power led to the creation of the now wholly independent branch of service, the U.S. Air Force, in 1947. Two years later, the various branches of military service were unified under the Department of Defense (except the Coast Guard, which is overseen by the U.S. Department of Homeland Security).

Since then, military pilots have played an integral role during the Cold War, Korean War, Vietnam War, Persian Gulf War, the wars in Afghanistan and Iraq, and countless smaller skirmishes and engagements, as well as in noncombat and peacekeeping situations.

THE JOB

Military pilots operate many different jet and propeller planes. Aircraft range from combat airplanes and helicopters, to supersonic fighters and bombers. In addition to actually flying aircraft, military pilots are also responsible for developing flight plans; checking weather reports; briefing and directing all crew members; and performing system operation checks to test the proper functioning of instrumentation, controls, and electronic and mechanical systems on the flight deck. They also are responsible for coordinating their takeoffs and landings with airplane dispatchers and air traffic controllers. At times, military pilots may be ordered to transport equipment and personnel, take reconnaissance photographs, spot and observe enemy positions, and patrol areas to carry out flight missions. After landing, military pilots must follow "afterlanding and shutdown" checklist procedures, and inform maintenance crews of any discrepancies or other problems noted during the flight. They must also present a written or oral flight report to their commanding officer.

There are several military piloting specialties. *Flight instructors* teach flight students how to fly via classroom training and inflight instruction. *Test pilots* play an important role in the testing and development of new aircraft and related technologies. They are employed by aerospace companies, the National Aeronautics and Space Administration (NASA), and the U.S. military, primarily the air force and the navy. Combining knowledge of flying with a background in aeronautical engineering, they test new models of planes and make sure they function properly. Test pilots are sometimes called *research pilots, research test pilots,* and *experimental test pilots.*

Detailed opportunities for military pilots in the five military branches are described below.

Although the army is best known for its land-based occupations, it also employs military pilots to serve in combat, rescue, and reconnaissance settings. Army pilots are classified under the warrant officer designation along with other skilled experts in nonaviation related fields.

The air force has the largest number of military pilots. These pilots work in a variety of specialty areas including bombers, airlifts, special operations, surveillance, and navigation. Specific job titles in this branch of the military include *air battle managers, airlift pilots and navigators, bomber pilots and navigators, fighter pilots and navigators, reconnaissance/surveillance/electronic warfare pilots and navigators, special operations pilots and navigators*, and *tanker pilots and navigators.*

Marine aviation officers provide air support for ground troops during battle. They also transport equipment and personnel to various destinations.

Pilots in the navy are called *naval flight officers.* Unlike other military pilots, they take off and land their airplanes on both land bases and aircraft carriers. Depending on their specialty, they receive advanced training in air-to-air combat, bombing, search and rescue, aircraft carrier qualifications, over-water navigation, and low-level flying. Naval flight officer specialties include *turboprop maritime propeller pilots,* who track submarines, conduct surveillance, and gather photographic intelligence, and *helicopter pilots,* who search for underwater supplies, deliver supplies and personnel, and participate in emergency search and rescue missions.

The U.S. Coast Guard is the only armed force in the United States with domestic law enforcement authority. Its aviators enforce federal laws and treaties and conduct military operations to safeguard the American homeland.

REQUIREMENTS

High School

You will need at least a high school degree in order to join the armed forces, and a college-preparatory curriculum is recommended. High school courses in science, mathematics, physics, computers, and physical education will be the most helpful. It's also a good idea to take a foreign language.

Postsecondary Training

A four-year college degree is usually required to become a military pilot. Courses in engineering, meteorology, computer science, aviation law, business management, and military science are especially

Rank and Military Branch By Occupation

Job Title	Rank	Military Branches
Airplane Pilots	Officer	Air Force, Army, Coast Guard, Marines, Navy
Helicopter Pilots	Officer	Air Force, Army, Coast Guard, Marines, Navy

Source: U.S. Department of Defense

helpful. Physical education courses will also be important, as your physical health and endurance levels will constantly be challenged in the military.

You can choose from several paths to get your postsecondary education. You may want to attend one of the four service academies: the U.S. Air Force Academy (for the air force), the U.S. Military Academy (for the army), the U.S. Naval Academy (for the navy and the marines), or the U.S. Coast Guard Academy (for the Coast Guard). Competition to enter these institutions is intense. You will need to have a very strong academic background, involvement in community activities, and leadership experiences. Most applicants also need a nomination from an authorized source, which is usually a member of the U.S. Congress. If you choose one of these four academies, you will graduate with a bachelor's degree. You are then required to spend a minimum of five years on active duty, beginning as a junior officer.

Another option is to attend a four-year school that has a Reserve Officers Training Corps (ROTC) program. Most state-supported colleges and universities have aviation programs, as do many private schools. Some schools focus solely on aviation education, such as Embry-Riddle Aeronautical University (http://www.erau.edu). Others such as the University of North Dakota (http://www.aero.und.edu/f1_Home/index.php), are well-known for their aviation and aerospace science programs.

Test pilots receive their training at civilian flight schools, such as the National Test Pilot School, or via military flight test schools (the U.S. Air Force Test Pilot School or the U.S. Naval Test Pilot School).

Each branch of the armed services has specific training requirements for its military pilots. Training in all branches will include flight simulation, classroom training, and basic flight instruction. For more information on specific requirements, contact a recruiter for the branch in which you are interested in entering.

Other Requirements

Stable physical and emotional health is essential for the aspiring pilot. Military pilots are expected to remain calm and levelheaded, no matter how stressful the situation. The physical requirements of

An air force pilot climbs the ladder to the cockpit of his F-16 Fighting Falcon as he prepares for a sortie from the Virginia Air National Guard Base in Sandstone, Virginia. *(Tech. Sergeant Ben Bloker, U.S. Air Force, U.S. Department of Defense)*

this profession are very strict—you must have 20/20 vision with or without glasses, good hearing, normal heart rate and blood pressure, and no physical handicaps that could hinder performance.

You should have quick decision-making skills and reflexes to be a successful pilot. Decisiveness, self-confidence, good communication skills, and the ability to work well under pressure are also important personality traits. You should maintain an adaptable and flexible lifestyle, as your orders, missions, and station may change at any time.

Although military pilot careers are available to both men and women, some combat positions are only open to men.

EXPLORING

Military recruiters often visit high schools, so be sure to take advantage of this opportunity to learn more about this field. Take a tour of a military base or an aircraft carrier if you get the chance. Talk with family and friends who have served in the armed forces to get advice and information.

To get a real feel for what it's like to be a military pilot, check out one of several air combat schools that exist throughout the country. Through such programs, you can experience the cockpit of a fighter plane alongside an instructor, and even experience "dogfighting" in the sky. Air Combat USA, which is one such program, operates out of 39 airports nationwide. See the For More Information section at the end of this article to learn more.

EMPLOYERS

Military pilots are employed by the U.S. government. There are approximately 16,000 airplane pilots and 6,500 helicopter pilots in the military.

STARTING OUT

Once you've decided to become a military pilot, you should contact a military recruiter. The recruiter will help answer questions and suggest different options. To start out in any branch of the military, you must pass medical and physical tests, the Armed Services Vocational Aptitude Battery exam, and basic training.

ADVANCEMENT

Each military branch has 10 officer grades (O-1 through O-10) and five warrant officer grades (W-1 through W-5). The higher the number is, the more advanced a person's rank is. The various branches of the military have somewhat different criteria for promoting individuals; in general, however, promotions depend on factors such as length of time served, demonstrated abilities, recommendations, and scores on written exams. Promotions become more and more competitive as people advance in rank.

Military pilots may train for different aircraft and missions. Eventually, they may advance to senior officer or command positions. Military pilots with superior skills and training may advance to the position of *astronaut*. Astronauts pilot the space shuttle on scientific and defense-related missions.

EARNINGS

The U.S. Congress sets the pay scales for the military after hearing recommendations from the president. The pay for equivalent grades is the same in all services (that is, anyone with a grade of O-4, for example, will have the same basic pay whether in the army, navy, marines, air force, or Coast Guard). In addition to basic pay, personnel who frequently and regularly participate in combat may earn hazardous duty pay. Other special allowances include special duty pay and foreign duty pay. Earnings start relatively low but increase on a fairly regular basis as individuals advance in rank. See the appendix at the end of this book for detailed information on pay scales for the U.S. military. When reviewing earnings, it is important to keep in mind that members of the military receive free housing, food, and health care—items that civilians typically pay for themselves.

Additional benefits for military personnel include uniform allowances, 30 days' paid vacation time per year, and the opportunity to retire after 20 years of service. Generally, those retiring will receive 40 percent of the average of the highest three years of their base pay. This amount rises incrementally, reaching 75 percent of the average of the highest three years of base pay after 30 years of service. All retirement provisions are subject to change, however, and you should verify them as well as current salary information before you enlist. Those who retire after 20 years of service are usually in their 40s and thus have plenty of time, as well as an accumulation of skills, with which to start a second career.

WORK ENVIRONMENT

The work environment for military pilots is rewarding, varied, and sometimes stressful. Pilots may be assigned to one or more air bases around the world. They may take off and land on aircraft carriers, at conventional airports, in desert conditions under fierce fire from the enemy, or in countless other settings. They may fly the same routes for extended periods of time, but no two flights are ever the same. Military pilots can expect excitement and the chance to see the world, but they are responsible for the safety and protection of others.

OUTLOOK

The outlook for military workers, including military pilots, is expected to be good through 2014, according to the U.S. Department of Labor. While political and economic conditions will have an influence on the military's duties and employment outlook, it is a fact that the country will always need military pilots, both for defense and to protect its interests and citizens around the world.

In the civilian sector, employment for pilots will grow about as fast as the average for all occupations through 2014, according to the U.S. Department of Labor.

FOR MORE INFORMATION

To read Looking for a Career Where the Sky Is the Limit?, *visit the ALPA's Web site.*

Air Line Pilots Association, International (ALPA)
1625 Massachusetts Avenue, NW
Washington, DC 20036-2212
http://www.alpa.org

For information on licensing, contact

Federal Aviation Administration
800 Independence Avenue, SW, Room 810
Washington, DC 20591-0001
Tel: 866-835-5322
http://www.faa.gov

For information on test pilots, contact

Society of Experimental Test Pilots
PO Box 986
Lancaster, CA 93584-0986
Tel: 661-942-9574
Email: Setp@setp.org
http://www.setp.org

To learn more about combat school experience, as well as pilot proficiency training, visit

Air Combat USA
http://www.aircombatusa.com

Read up on military news and developments by visiting

Defense-Aerospace.com
http://www.defense-aerospace.com

To get information on specific branches of the military, check out this site, which is the home of ArmyTimes.com, NavyTimes.com, AirForceTimes.com, and MarineCorpsTimes.com:

Military Times
http://www.militarytimes.com

Visit the following Web sites for information on test pilot schools:

National Test Pilot School
http://www.ntps.com

U.S. Air Force Test Pilot School
http://www.edwards.af.mil

U.S. Naval Test Pilot School
http://www.usntps.navy.mil

If you're thinking of joining the armed forces, take a look at this site, which guides students and parents through the decision-making process:

Today's Military
http://www.todaysmilitary.com

For information on becoming a military pilot, contact

United States Air Force
http://www.airforce.com

United States Army
http://www.goarmy.com

United States Coast Guard
http://www.gocoastguard.com/officerindex.html

United States Marine Corps
http://www.marines.com

United States Navy
http://www.navy.com/officer/aviation

Military Recruiters

QUICK FACTS

School Subjects
Government
Psychology
Speech

Personal Skills
Communication/ideas
Leadership/management

Work Environment
Primarily indoors
Primarily multiple locations

Minimum Education Level
Varies by career specialty

Salary Range
$15,617 to $30,618 to $66,154 (enlisted personnel)
$25,358 to $70,877 to $174,103 (officers)

Outlook
About as fast as the average

DOT
166

GOE
13.02.01

O*NET-SOC
13-1071.02

OVERVIEW

Military recruiters, sometimes known as *recruiting specialists,* provide information regarding service, training, and career opportunities to people interested in joining a branch of the military. They represent the military at job fairs and career programs, and with community and school groups. Their duties include interviewing, screening, testing, and counseling possible candidates. *Recruiting managers* direct recruiters, plan recruiting programs, and prepare reports for commanders that detail the progress of their recruiting programs. Opportunities are available in all five military branches.

HISTORY

As competition has increased in the business world in the past few decades, a need has developed for professionals who can recruit top managers or workers with unique job training that will help their employers gain a competitive advantage. Companies began increasingly turning to a third party for their employment needs—the recruiter.

Military recruiters have encouraged people to join the armed forces since the Revolutionary War. During World War I, the iconic "I Want You for the U.S. Army" poster made its first appearance. The poster shows Uncle Sam, a man with a white beard and hat, dressed in red, white, and blue (a personification of the United States), pointing from the poster at readers in an attempt to induce them to join the military. The poster was also used during World War II, and is occasionally referenced in military recruiting ads today.

Recruiting has gone high tech since those days. Today, recruiters use print and broadcast advertisements, emails and text messages, and Web sites to reach potential recruits. They also try to reach potential military enlistees by visiting high schools, staffing booths at college fairs, traveling to college campuses, and working at recruiting centers.

THE JOB

Men and women consider joining the military for various reasons. For some, it's the honor of serving their country; some view the military as a way to earn money to pay for a college education. Others seek the valuable training and work experience to help them establish careers in the civilian workforce. Still others hope to make lifelong careers with the military branch of their choice. The first step for many of these career paths is a visit with a military recruiter.

Military recruiters provide information to civilians interested in a possible military career. Each branch of the military—the army, air force, Coast Guard, marines, and navy—has their own recruiters working on its behalf. The army also employs civilian recruiters, as mandated by the Defense Appropriations Act of 2001. While the two companies currently contracted to manage recruiting personnel are considered civilian, many of those hired to conduct recruitment duties have had previous distinguished service in the army.

Some military recruiters are based at recruitment offices located throughout the United States. They meet with candidates by appointment or on a walk-in basis and provide information and answer any questions they may have. Many recruiters also travel to area high schools and college campuses to make presentations to interested students. Information regarding military careers is also presented at job fairs, career day programs, and community groups. Recruiters may hand out pamphlets, give testimonials, or present a slideshow regarding the many opportunities available in the military. They give an honest account of what to expect during a tour of duty, both positive and negative, as well as the financial and career benefits of military service.

Military recruiters conduct a preliminary interview with an interested candidate. During this time the recruiter can assess whether the candidate is qualified to join. There are standards set by the U.S. Department of Defense, along with each military branch, which determines qualification. Such factors include age (you must be of age to serve), citizenship or immigration status, level of education, criminal or drug abuse history, physical and mental health, and level of education. Many candidates, while interested in a military career, may be unsure of what branch in which to serve. They often visit with recruiters from every branch before deciding with whom to enlist.

Once the candidate decides on a branch of the military, the recruiter presents the particulars of service. Issues such as length of service, basic training, leave time, housing options, medical benefits, education benefits, and financial matters are discussed in depth. Recruiters often meet with candidates more than once to answer questions, explain details at length, or assuage any concerns. They may be asked to meet with a candidate's parents, guardians, or family members to answer their queries or concerns.

Military recruiters will ask candidates to take the Armed Forces Vocational Aptitude Battery test. This computerized test covers topics such as word knowledge and comprehension, and mathematical knowledge and reasoning. Scores from this test may help determine in what capacity the candidate can best serve the military.

The enlistment process continues with the recruiter drawing up forms for the candidate which specifies his or her commitment to join the military. The recruiter often helps the candidate gather needed paperwork and documents to complete the application. While the recruiter can schedule and even provide transportation to a Military Entrance Processing Station, they cannot accompany the candidate through this part of the process.

Recruiters can advise candidates regarding job opportunities or specifics but they cannot guarantee a particular job description. Recruiters working for the army, including the Army Reserves and the Army National Guard however, have access to the Future Soldier Remote Reservations System, which may give prequalified candidates the option of requesting a particular job—only if that position is available at the time of their processing. The remaining branches use job counselors to help candidates match their qualifications with available positions.

REQUIREMENTS

High School

To prepare for a career as a recruiter, you should take business, speech, English, and mathematics classes in high school. Psychology and sociology courses will teach you how to recognize personality characteristics that may be key in helping you determine which recruits would best fit a position.

Postsecondary Training

In the civilian sector, most recruiters have a bachelor's degree in human resources management or a business degree with a concentration in human resources management.

Rank and Military Branch By Occupation

Job Title	Rank	Military Branches
Recruiting Managers	Officer	Air Force, Coast Guard, Marines, Navy
Recruiting Specialists	Enlisted	Air Force, Army, Coast Guard, Marines, Navy

Source: U.S. Department of Defense

Military recruiters train via classroom instruction, on-the-job experience, and advanced course work. Classes cover recruiting methods, public speaking techniques, interviewing techniques, and community relations practices.

Visit the U.S. Department of Defense's Web site, http://www.todaysmilitary.com, for more on training for those interested in becoming military recruiters.

Other Requirements

Since much of the work is done face to face, recruiters need to be comfortable speaking in front of people. Public speaking skills are necessary whether addressing a gathering of students or community groups, or when interviewing a single candidate.

It is equally important to put forth a good first impression. Candidates, especially if they are unsure of which area to join, may be influenced by the rapport they develop with their respective military recruiter. Appearing too brash, insensitive, or indifferent may cause a potential candidate to look to another branch of the military, or reconsider his or her desire to serve.

Patience is another important personal skill. Recruiters realize that the decision to serve is an important one, and should give potential candidates the information, counseling, and time to make the right choice.

EXPLORING

Familiarize yourself with business practices by joining or starting a business club at your school. Being a part of a speech or debate team is a great way to develop excellent speaking skills, which are necessary in this field. Hold mock interviews with family or friends. Professional associations, such as the International Association of

Steps in the Enlistment Process

If you are interested in joining the military, it is a good idea to learn more about the enlistment process before you make a commitment. There are three main stages in the enlistment process: Prequalification, Screening, and the Military Entrance Processing Station. Read on to learn more about these steps.

Prequalification

Once you talk to your recruiter and you both decide that you are a good candidate for the military, the recruiter will need to gather some basic information from you. This includes your Social Security number, a copy of your driver's license, a copy of your birth certificate, and proof of high school graduation (or equivalent degree such as the GED). You may also be asked to complete a nonbinding application form.

Screening

Next, you will be screened for your "physical status, certain moral beliefs and facts, and aptitudes." To check your physical status, the recruiter will ask you to provide a complete record of your medical history. He or she will also check to see if your height and weight meets the standards of the military branch you are interested in joining.

In regard to moral beliefs, the recruiter will ask you if you have ever used illegal drugs and/or have a police record. He or she will also ask you if you are a conscientious objector, "an individual who has a fixed, firm, and sincere objection to participation in war in any form or to the performance of military service because of religious training and beliefs."

Finally, the recruiter will test your aptitudes. If you haven't taken the Armed Services Vocational Aptitude Battery (ASVAB) test (http://www.todaysmilitary.com/app/tm/nextsteps/asvab), the recruiter may ask you to take a preliminary aptitude screening test called the Entrance Screening Test (EST). This tests tells the recruiter if you will achieve an acceptable score on the ASVAB. If you score well on the EST, your recruiter will most likely have you take the ASVAB to further gauge your skills and aptitudes.

Military Entrance Processing Station

If you pass the Prequalification and Screening phases and still want to join the military, your next step is to go to a Military Entrance Processing Station. There, you will take the ASVAB (if you haven't already done so). You will also undergo a physical, participate in a pre-enlistment interview, and finally take the Oath of Enlistment. Visit http://www.mepcom.army.mil for detailed information on this step in the process.

Source: U.S. Department of Defense

Corporate and Professional Recruitment, are also good sources of information. Visit this association's Web site at http://www.iacpr.org to learn more.

Others ways to learn about this career include reading books and magazines about military recruiters and asking your guidance counselor or teacher to arrange an information interview with a civilian or military recruiter.

To learn more about career opportunities in the military, visit the Web sites listed at the end of this article.

EMPLOYERS

The U.S. government and two private companies contracted by the U.S. Department of Defense employ approximately 4,550 military recruiters.

STARTING OUT

Work as a military recruiter is not typically an entry-level job. Most recruiters enter the field after serving in the military for at least a few years.

ADVANCEMENT

Most recruiters only stay in the field for a few years before moving on to positions in personnel or administration in the military or in private industry. Recruiters who decide to make a career of military recruiting may advance to supervise one or more recruiting offices or be assigned to senior management positions.

EARNINGS

The U.S. Congress sets the pay scales for the military after hearing recommendations from the president. The pay for equivalent grades is the same in all services (that is, anyone with a grade of E-4, for example, will have the same basic pay whether in the army, navy, marines, air force, or Coast Guard). In addition to basic pay, personnel who frequently and regularly participate in combat may earn hazardous duty pay. Other special allowances include special duty pay and foreign duty pay. Earnings start relatively low but increase on a fairly regular basis as individuals advance in rank. See the appendix at the end of this book for detailed information on pay scales for the U.S. military. When reviewing earnings, it is important to keep in mind that members of the military receive free housing, food, and health care—items that civilians typically pay for themselves.

Additional benefits for military personnel include uniform allowances, 30 days' paid vacation time per year, and the opportunity to retire after 20 years of service. Generally, those retiring will receive 40 percent of the average of the highest three years of their base pay. This amount rises incrementally, reaching 75 percent of the average of the highest three years of base pay after 30 years of service. All retirement provisions are subject to change, however, and you should verify them as well as current salary information before you enlist. Those who retire after 20 years of service are usually in their 40s and thus have plenty of time, as well as an accumulation of skills, with which to start a second career.

WORK ENVIRONMENT

Much of a recruiter's work is done in an office setting, such as a local recruiting office. They often travel to different locations, especially when making presentations at high schools or college campuses, or staffing a booth at a local job fair. Recruiters can expect to work some evening or weekend hours to accommodate candidates' needs, or the needs of their families.

OUTLOOK

Employment in the armed forces is expected to grow about as fast as the average for all occupations through 2014, according to the U.S. Department of Labor. When the economy is good and/or during times of war, more people pursue employment in the civilian workforce, which creates additional opportunities in the military. With the U.S. military involved in several international conflicts, most significantly in Iraq and Afghanistan, demand should continue to be strong for military workers. Military recruiters will be in steady demand over the next decade as the U.S. military tries to meet recruiting goals in all five military branches. More than 340,000 men and women enlist in the U.S. military each year.

The civilian executive search industry should have a good future. Potential clients include not only large international corporations but also universities, the government, and smaller businesses. Smaller operations are aware that having a solid executive or administrator may make the difference between turning a profit or not being in business at all. Many times, search firm services are used to conduct industry research or to scope out the competition. Executive search firms now specialize in many fields of employment—health care, engineering, or accounting, for example.

FOR MORE INFORMATION

For industry information, contact

National Association of Executive Recruiters
1901 North Roselle Road, Suite 920
Schaumburg, IL 60195-3187
Tel: 847-885-1453
Email: naerinfo@naer.org
http://www.naer.org

To get information on specific branches of the military, check out this site, which is the home of ArmyTimes.com, NavyTimes.com, AirForceTimes.com, and MarineCorpsTimes.com:

Military City
http://www.militarytimes.com

If you're thinking of joining the armed forces, take a look at this site, which guides students and parents through the decision-making process:

Today's Military
http://www.todaysmilitary.com

For information on military careers, contact

United States Air Force
http://www.airforce.com

United States Army
http://www.goarmy.com

United States Coast Guard
http://www.gocoastguard.com

United States Marine Corps
http://www.marines.com

United States Navy
http://www.navy.com

Naval and Maritime Operations Occupations

QUICK FACTS

School Subjects
Computer science
Technical/shop

Personal Skills
Following instructions
Mechanical/manipulative

Work Environment
Indoors and outdoors
Primarily multiple locations

Minimum Education Level
Varies by career specialty

Salary Range
$15,617 to $30,618 to $66,154 (enlisted personnel)
$25,358 to $70,877 to $174,103 (officers)

Outlook
About as fast as the average

DOT
911

GOE
07.04.01

O*NET-SOC
53-5011.00, 53-5011.01, 53-5011.02, 53-5021.00, 53-5021.01, 53-5021.02, 53-5021.03, 53-5031.00

OVERVIEW

Officers and crewmembers specializing in naval and maritime operations use vessels to patrol harbors, transport troops and supplies, and conduct search and rescue missions. They are responsible for the navigation and operation of many different water vessels ranging from small tugboats and submarines to large ships. The U.S. Army, Coast Guard, Marines, and Navy offer various employment opportunities in naval and marine operations at both the enlisted and officer levels.

HISTORY

Merchant shipping is an old industry that developed out of the need and desire to trade and travel. In the early days of the American colonies, commercial shipping was very important. Deep-water rivers and channels provided perfect launching sites for water vessels built by craftspeople who emigrated from other countries. Between 1800 and 1840, U.S. ships carried more than 80 percent of the country's commerce with other nations. The first steamship crossed the Atlantic Ocean in 1819, and large iron ships began to be built in the mid-1800s. With the mass production of these iron ships, trade increased and Great Britain dominated the industry through the end of the century. Most U.S. trade was carried by foreign ships.

The merchant marine has always been a private industry, but the government has relied on it to help in a military capacity during times of war. The maritime industry benefits during wartime because the country's defense department contracts shipbuilders and merchant mariners to serve as auxiliaries to the military. In fact, the end of the world wars caused a depression in U.S. merchant shipping. During World War II, the U.S. Merchant Marine Academy was established at Kings Point, New York, to offer training for merchant officers (it didn't admit women until 1974). There are now six maritime academies in the United States.

The oldest continuous seagoing service in the United States, the Coast Guard, was established in 1790 to combat smuggling. In contrast, the first American marine units were attached to the army at the time of its creation; these units then were made an independent part of the navy when it was officially established in 1798. The Marine Corps was considered part of the navy until 1834, when it established itself as both a land and sea defense force, thereby becoming its own military branch.

THE JOB

The following paragraphs describe some of the more popular career options for people interested in naval and maritime careers in the military.

Quartermasters and boat operators operate and navigate small watercrafts. They steer tugboats when towing large ships out to sea, or docking supply barges. During times of war, they maneuver landing crafts during an amphibious assault. Other duties include operating navigation systems and ship to shore communication, boat and dock maintenance, and data input for the ship's logs.

Seamen, a position for men and women, perform many duties crucial to the daily operation of ships, boats, submarines, and other maritime vessels. Seamen operate and care for the ship's equipment such as rigging, hoists, cranes, and gangplanks. Maintenance duties include cleaning, painting, and repairing the deck of a ship, or the interior and exterior of a submarine. They are also in charge of containing any fires or other emergencies that happen on board. Some seamen may be assigned to navigation or security details. Seamen enlisted in the navy are further named according to the department they work in—for example, those working in the ship's hull or engineering department are referred to as *firemen*.

Ship and submarine officers manage the crew and activities of different departments within each military vessel. Ship officers help the captain command and navigate ships while at sea, plan and implement emergency drills, and lead target practice training sessions. They may also direct search and rescue missions. Ship officers may have specialized duties according to the branch they serve. For example, during times of national crisis *Coast Guard ship officers* may be assigned to patrol coastal waters. *Naval submarine officers* may be responsible for operating the command and control system on a nuclear submarine.

Military ships rely on enormous engines to power their vessels, including providing electricity and heat. *Ship engineers* are responsible for the daily operation, maintenance, and repair of these nuclear- or diesel-powered engines. Their duties include directing crews assigned to the engine room and inspecting and maintaining heating, cooling, communication, and propulsion systems.

REQUIREMENTS

High School

Mathematics and physics courses provide good training for a number of nautical activities. Computer science will prepare you for the increasing use of high technology at sea, and physical education will get you in shape for the sometimes strenuous work on a ship.

Postsecondary Training

In the civilian sector, a good way to fulfill many requirements and also learn about the various types of shipboard work is to attend a maritime school, such as Massachusetts Maritime Academy (http://www.maritime.edu), U.S. Merchant Marine Academy (http://www.usmma.edu), or Maine Maritime Academy (http://www.mainemaritime.edu).

Enlisted personnel (quartermasters, boat operators, and seamen) train via classroom instruction and on-the-job experience. Quartermasters and boat operators learn via classroom instruction and practice experience in boat operations. Classes cover boat handling procedures, navigational mathematics, and the use of compasses, charts, radar, and other navigational aids. Seamen learn how to operate equipment through on-the-job training.

Officers (ship engineers, ship and submarine officers) must enter the military with at least a bachelor's degree. Once they join the military, they receive additional specialized training via classroom instruction, on-the-job experience, and advanced course work. Ship engineers take classes that teach them how to inspect and maintain marine engines, fuel systems, and electrical systems and maintain and operate power plants and related machinery. Training for ship engineers who specialize in nuclear technology lasts the longest. Ship and submarine officers train via classroom instruction and practical hands-on experience in one of the following specialties: administration, air, communications, deck, engineering, operations, supply, or weapons.

Visit the U.S. Department of Defense's Web site, http://www.todaysmilitary.com, for more on military training in naval and maritime operations careers.

Other Requirements

Officers and crew must be team players, and able to perform their duties even under extreme pressure. The military stresses physical health and conditioning; enlisted naval and marine operations crew members often have duties that require stamina for physical work.

It is also important to be interested in boating and be comfortable being at sea. Officers and crew often have to endure long stretches of time at sea, especially when sent on missions abroad.

EXPLORING

Assuming you are already accustomed to being on a boat, there are very few opportunities to explore this field before actually enrolling in a maritime program, applying at union halls or shipping companies, or joining the military. If near a port, you could visit a vessel in port by contacting a steamship company. Visiting coastal ports (e.g., in Maine or California) is a good idea.

Others ways to learn about any of the jobs listed in this article include reading books and magazines about these careers, visiting Web sites of professional associations, and asking your guidance counselor or teacher to arrange an information interview with a worker in the field.

To learn more about career opportunities in the military, visit the Web sites listed at the end of this article.

EMPLOYERS

The U.S. government employs the military. No specific employment statistics are available for military workers in naval and maritime operations. Overall, 1.4 million men and women are on active duty and another 1.2 million volunteers serve in the Nation Guard and Reserve forces.

STARTING OUT

If you are interested in learning more about careers in the military, you should contact a military recruiter. Visit the Web sites listed at the end of this article to locate a recruiting office near you. To start out in any branch, you will need to pass physical and medical tests, the Armed Services Vocational Aptitude Battery exam, and basic training.

Rank and Military Branch By Occupation

Job Title	Rank	Military Branches
Quartermasters and Boat Operators	Enlisted	Army, Coast Guard, Marines, Navy
Seamen	Enlisted	Coast Guard, Navy
Ship and Submarine Officers	Officer	Army, Coast Guard, Navy
Ship Engineers	Officer	Army, Coast Guard, Navy

Source: U.S. Department of Defense

ADVANCEMENT

Each military branch has nine enlisted grades (E-1 through E-9) and 10 officer grades (O-1 through O-10). The higher the number is, the more advanced a person's rank is. The various branches of the military have somewhat different criteria for promoting individuals; in general, however, promotions depend on factors such as length of time served, demonstrated abilities, recommendations, and scores on written exams. Promotions become more and more competitive as people advance in rank. On average, a diligent enlisted person can expect to earn one of the middle noncommissioned or petty officer rankings (E-4 through E-6); some officers can expect to reach lieutenant colonel or commander (O-5). Outstanding individuals may be able to advance beyond these levels.

EARNINGS

The U.S. Congress sets the pay scales for the military after hearing recommendations from the president. The pay for equivalent grades is the same in all services (that is, anyone with a grade of E-4, for example, will have the same basic pay whether in the army, navy, marines, air force, or Coast Guard). In addition to basic pay, personnel who frequently and regularly participate in combat may earn hazardous duty pay. Other special allowances include special duty pay and foreign duty pay. Earnings start relatively low but increase on a fairly regular basis as individuals advance in rank. See the appendix at the end of this book for detailed information on pay scales for the U.S. military. When reviewing earnings, it is important to keep in mind that members of the military receive free housing, food, and health care—items that civilians typically pay for themselves.

Additional benefits for military personnel include uniform allowances, 30 days' paid vacation time per year, and the opportunity to retire after 20 years of service. Generally, those retiring will receive 40 percent of the average of the highest three years of their base pay. This amount rises incrementally, reaching 75 percent of the average of the highest three years of base pay after 30 years of service. All retirement provisions are subject to change, however, and you should verify them as well as current salary information before you enlist. Those who retire after 20 years of service are usually in their 40s and thus have plenty of time, as well as an accumulation of skills, with which to start a second career.

WORK ENVIRONMENT

Much of the work is done in small, confined spaces, which may be uncomfortable at times. If assigned to the engine room, for example, be prepared to encounter high temperatures and extreme noise levels. Seamen, quartermasters, and boat operators often have duties that must be done above deck or ashore, regardless of the weather. Those working for the military are sometimes involved in dangerous situations, especially during times of war or conflict.

OUTLOOK

Employment in the armed forces is expected to grow about as fast as the average for all occupations through 2014, according to the U.S. Department of Labor. When the economy is good and/or during times of war, more people pursue employment in the civilian workforce, which creates additional opportunities in the military. With the U.S. military involved in several international conflicts, most significantly in Iraq and Afghanistan, demand should continue to be strong for military workers, including those in naval and maritime operations.

In the civilian sector, the employment outlook for merchant marine personnel is not very good, mainly because of foreign competition and changes in federal policy. Cargo rates and wages paid to U.S. merchant mariners are the highest in the world, but this keeps the industry small because shippers can send goods on cheaper foreign vessels. However, new international regulations have raised shipping standards with respect to safety, training, and working conditions, so less competition is expected from foreign vessels that will have to pay higher insurance rates for ships that do not meet the standards. The U.S. Department of Labor projects that overall, employment in water transportation occupations will grow more slowly than the average for all occupations through 2014.

FOR MORE INFORMATION

For information on union membership, contact

International Organization of Masters, Mates, and Pilots
700 Maritime Boulevard, Suite B
Linthicum Heights, MD 21090-1941
http://www.bridgedeck.com

For publications, statistics, and news, contact

U.S. Maritime Administration
U.S. Department of Transportation
1200 New Jersey Avenue, SE, Second Floor, West Building
Washington, DC 20590-0001
http://www.marad.dot.gov

For information on academic training, contact

U.S. Merchant Marine Academy
300 Steamboat Road
Kings Point, NY 11024-1634
Tel: 516-773-5000
http://www.usmma.edu

To get information on specific branches of the military, check out this site, which is the home of ArmyTimes.com, NavyTimes.com, AirForceTimes.com, and MarineCorpsTimes.com:

Military City
http://www.militarytimes.com

If you're thinking of joining the armed forces, take a look at this site, which guides students and parents through the decision-making process:

Today's Military
http://www.todaysmilitary.com

For information on military careers, contact

United States Army
http://www.goarmy.com

United States Coast Guard
http://www.gocoastguard.com

United States Marine Corps
http://www.marines.com

United States Navy
http://www.navy.com

Personal and Culinary Services Occupations

OVERVIEW

The military relies on workers in personal and culinary services to keep morale high and prepare nutritious food for soldiers and their families at military bases, on ships and submarines, and in the field. Opportunities are available for both enlisted personnel and officers in the U.S. Air Force, Army, Coast Guard, Marines, and Navy. There were 32,610 soldiers employed in support service occupations in 2005.

QUICK FACTS

School Subjects
Family and consumer science
Mathematics

Personal Skills
Artistic
Following instructions

Work Environment
Primarily indoors
Primarily multiple locations

Minimum Education Level
Varies by career specialty

Salary Range
$15,617 to $30,618 to $66,154 (enlisted personnel)
$25,358 to $70,877 to $174,103 (officers)

Outlook
About as fast as the average

DOT
157, 195, 311, 313

GOE
11.01.01, 11.02.01, 11.05.01, 11.01.02

O*NET-SOC
11-9051.00, 35-1011.00, 35-1012.00, 35-2011.00, 35-2012.00, 35-2014.00, 35-2015.00, 35-2021.00, 35-3022.00, 35-3031.00, 35-9011.00, 39-9032.00

HISTORY

In ancient and medieval times, inns were established along main highways to provide food and lodging for travelers. Usually, the innkeeper and his or her family, with perhaps a few servants, were able to look after all the needs of travelers. Restaurants as we know them today hardly existed. Wealthy people did almost all their entertaining in their own homes, where they had large staffs of servants to wait on their guests.

Improved roads and transportation methods in the 18th and 19th centuries led to an increase in travel for both business and pleasure. Inns near large cities, no longer merely havens for weary travelers, became pleasant destinations for day excursions into the country. The rise of an urban middle class created a demand for restaurants where people could enjoy good food and socialize in a convivial atmosphere. More and more waiters were needed to serve the growing number of customers. In the great hotels and restaurants of Europe in the 19th century, the presentation of elegantly prepared food in a polished and gracious manner was raised to a high art.

In the United States, the increasing ease and speed of travel has contributed to a very mobile population, which has created a greater demand for commercial food service. People eat at restaurants and fast food establishments more and more. Today, the food service industry is one of the largest and most active sectors of the nation's economy.

In the military, culinary services professionals have been in demand ever since America's first fighting force, the Continental Army, was formed in 1775 to battle the British during the Revolutionary War. Workers were needed to purchase and prepare food for soldiers during training and conflict. These foods ranged from items such as flapjacks, cooked meat, and hasty pudding during training to a ration of bread, salted beef or pork, and dry beans or peas during wartime. The need to obtain and prepare quality food and beverages for military personnel has not changed during more than 230 years of war and peacetime operations, but the methods and procedures have. Today, military culinary services professionals are preparing food in mess halls at military bases throughout the United States, on naval vessels and submarines patrolling the Indian Ocean, in kitchens and mess halls in Iraq and Afghanistan, and in countless other places throughout the world where the military serves.

THE JOB

The military relies on food service professionals to prepare nutritious meals for its more than 1.4 million active duty personnel, as well as those in the National Guard and other military entities. The following paragraphs detail popular career paths.

Food service managers operate mess halls and any other facilities that serve food (including mess halls in combat zones). They procure a variety of food; select the menu, attend to the staffing and equipment needs of kitchens, dining halls, and meat-cutting plants; help prepare the food; and, most importantly, ensure that health and sanitation standards are met.

Food service specialists order, receive, inspect, prepare, and serve food and beverages. They also clean, stoves, ovens, pots and pans, mixers, utensils, and any other equipment that is used to prepare and serve food and beverages. Food service specialists work in dining halls, field kitchens, hospitals, or aboard ships.

Recreation, welfare, and morale specialists provide recreational support services to members of the military and their families, with the goal of enhancing their general well-being. They organize recreational events such as sport programs, arts and crafts activities, and social gatherings. They must determine what types of activities

Rank and Military Branch By Occupation

Job Title	Rank	Military Branches
Food Service Managers	Officer	Air Force, Army, Coast Guard, Marines, Navy
Food Service Specialists	Enlisted	Air Force, Army, Coast Guard, Marines, Navy
Recreation, Welfare, and Morale Specialists	Enlisted	Contact the various military branches for more information on employment opportunities in these professions.

Source: U.S. Department of Defense

military members and their families are most interested in, then manage the materials, equipment, and facilities involved.

REQUIREMENTS

High School

If you are interested in becoming a food service manager or specialist, your high school education should include classes in family and consumer science and health. These courses will teach you about nutrition, food preparation, and food storage. Math classes are also recommended; in this line of work you must be comfortable working with fractions, multiplying, and dividing. Aspiring food service managers should also take business courses.

If you want to become a recreation, welfare, or morale specialist, take classes in psychology, speech, business, and physical education.

Postsecondary Training

Educational requirements in the civilian sector vary by career. For example, food service managers typically earn a bachelor's degree in restaurant and hotel management or institutional food service management. Some individuals qualify for management training by earning an associate's degree or other formal award below the bachelor's degree level. Food service specialists usually receive on-the-job training. Recreation workers (known as recreation, welfare, and morale specialists in the military) typically have bachelor's degrees in parks and recreation management, leisure studies, fitness management, or related disciplines. A degree in any liberal arts field may be sufficient if the person's education includes courses relevant to recreation work.

In the military, enlisted personnel (food service specialists and recreation, welfare, and morale specialists) train via classroom instruction and advanced course work. Course work varies by specialty.

As officers in the military, food service managers must enter the armed forces with at least a bachelor's degree. Once they join the military, they receive additional classroom training that covers topics such as food service operations and management, nutritional meal planning, and managing resources and supplies.

Visit the U.S. Department of Defense's Web site, http://www.todaysmilitary.com, for more on military training for those interested in personal and culinary services careers.

Other Requirements

Personal requirements for workers in these occupations vary by specialty. For example, food service managers must be familiar with the various operations of food preparation, service operations, sanitary regulations, and financial functions. They must also have good organizational skills and be able to manage people. Food service specialists must possess strong physical stamina, because the work requires many long hours of standing and walking. They also must be neat and clean in their personal hygiene. Recreation, welfare, and morale specialists must have good people skills, be interested in recreational activities, have excellent organizational skills, and enjoy helping and teaching others.

Visit the U.S. Department of Defense's Web site, http://www.todaysmilitary.com, for more on personal requirements for workers in these careers.

EXPLORING

Explore this work by getting part-time or summer work as a dining room attendant, counter worker, waiter, cook, or assistant manager at a restaurant, grill, or coffee shop. Volunteer opportunities that combine some type of food service and interaction with the public may also be available in your area. Meals on Wheels, shelters serving meals, and catering services are all sources to consult for volunteering opportunities.

Young people interested in the field of recreation should obtain related work experience as part-time or summer workers or volunteers in recreation departments, neighborhood centers, camps, and other organizations.

Other ways to learn about any of the jobs listed in this article include reading books and magazines about food service professionals and recreation workers, visiting Web sites of professional associations, arranging a tour of a restaurant kitchen or recreation center, and asking your guidance counselor or teacher to arrange an information interview with a worker in the field.

To learn more about career opportunities in the military, visit the Web sites listed at the end of this article.

EMPLOYERS

The U.S. government employs the military. In 2005, 32,610 soldiers were employed in support service occupations: 2,497 individuals served in the air force; 14,963 in the army; 1,146 in the Coast Guard; 2,302 in the marines; and 11,702 in the navy.

STARTING OUT

A military recruiter is the person to contact if you want to enter the armed forces. Visit the Web sites listed at the end of this article to locate a recruiting office near you. To start out in any branch, you will need to pass physical and medical tests, the Armed Services Vocational Aptitude Battery exam, and basic training.

ADVANCEMENT

Each military branch has nine enlisted grades (E-1 through E-9) and 10 officer grades (O-1 through O-10). The higher the number is, the more advanced a person's rank is. The various branches of the military have somewhat different criteria for promoting individuals; in general, however, promotions depend on factors such as length of time served, demonstrated abilities, recommendations, and scores on written exams. Promotions become more and more competitive as people advance in rank. On average, a diligent enlisted person can expect to earn one of the middle noncommissioned or petty officer rankings (E-4 through E-6); some officers can expect to reach lieutenant colonel or commander (O-5). Outstanding individuals may be able to advance beyond these levels.

EARNINGS

The U.S. Congress sets the pay scales for the military after hearing recommendations from the president. The pay for equivalent grades is the same in all services (that is, anyone with a grade of E-4, for example, will have the same basic pay whether in the army, navy, marines, air force, or Coast Guard). In addition to basic pay, personnel who frequently and regularly participate in combat may earn hazardous duty pay. Other special allowances include special duty pay and foreign duty pay. Earnings start relatively low but increase on a fairly regular basis as individuals advance in rank. See the appendix at

An air force chef cooks burgers on a grill at the Red River Dining Facility at Barksdale Air Force Base in Louisiana. *(Airman 1st Class Trina R. Flannagan, U.S. Air Force, U.S. Department of Defense)*

the end of this book for detailed information on pay scales for the U.S. military. When reviewing earnings, it is important to keep in mind that members of the military receive free housing, food, and health care—items that civilians typically pay for themselves.

Additional benefits for military personnel include uniform allowances, 30 days' paid vacation time per year, and the opportunity to retire after 20 years of service. Generally, those retiring will receive 40 percent of the average of the highest three years of their base pay. This amount rises incrementally, reaching 75 percent of the average of the highest three years of base pay after 30 years of service. All retirement provisions are subject to change, however, and you should verify them as well as current salary information before you enlist. Those who retire after 20 years of service are usually in their 40s and thus have plenty of time, as well as an accumulation of skills, with which to start a second career.

WORK ENVIRONMENT

Food service professionals work in food service facilities, kitchens, and dining facilities on land (at military installations and in tents in the field) and on ships and submarines. They are subject to certain work hazards. These may include burns from heat and steam; cuts and injuries from knives, glassware, and other equipment; and sometimes hard falls from rushing on slippery floors. The job also requires lifting heavy trays of food, dishes, and glassware, as well as a great deal of bending and stooping. In some cases, employees may work near steam tables or hot ovens.

Recreation, welfare, and morale specialists typically work in offices when organizing events, and outdoors when conducting recreational activities.

OUTLOOK

Employment in the armed forces is expected to grow about as fast as the average for all occupations through 2014, according to the U.S. Department of Labor. When the economy is good and/or during times of war, more people pursue employment in the civilian workforce, which creates additional opportunities in the military. With the U.S. military currently involved in several international conflicts, most significantly in Iraq and Afghanistan, demand should continue to be strong for military workers.

Military food service workers should continue to have excellent opportunities. It is important that soldiers receive nutritious meals to keep them performing at top health and efficiency. And with millions of soldiers needing to be fed daily, there will always be opportunities for workers in this field.

Recreation, welfare, and morale specialists will also have good opportunities. They will be needed to help military personnel and their families reduce stress in the field and on military bases.

According to the U.S. Department of Labor, employment for civilian workers in food service should grow about as fast as the average for all occupations through 2014. Employment for recreation workers is also predicted to grow about as fast as the average.

FOR MORE INFORMATION

For information regarding industry trends, accredited institutions, and conventions, contact

American Association for Physical Activity and Recreation
1900 Association Drive
Reston, VA 20191-1598
Tel: 703-476-3400
http://www.aahperd.org/aapar

For information on job opportunities and accredited education programs, contact

International Council on Hotel, Restaurant, and Institutional Education
2810 North Parham, Suite 230
Richmond, VA 23294-4422

Tel: 804-346-4800
Email: info@chrie.org
http://chrie.org

For information on education, scholarships, and careers, contact
National Restaurant Association Educational Foundation
175 West Jackson Boulevard, Suite 1500
Chicago, IL 60604-2814
Tel: 800-765-2122
Email: info@nraef.org
http://www.nraef.org

To get information on specific branches of the military, check out this site, which is the home of ArmyTimes.com, NavyTimes.com, AirForceTimes.com, and MarineCorpsTimes.com:
Military City
http://www.militarytimes.com

If you're thinking of joining the armed forces, take a look at this site, which guides students and parents through the decision-making process:
Today's Military
http://www.todaysmilitary.com

For information on military careers, contact
United States Air Force
http://www.airforce.com

United States Army
http://www.goarmy.com

United States Coast Guard
http://www.gocoastguard.com

United States Marine Corps
http://www.marines.com

United States Navy
http://www.navy.com

Transportation, Supply, and Logistics Occupations

OVERVIEW

The military relies on *transportation, supply, and logistics workers* for the quick and efficient transport of troops, equipment, and supplies anywhere in the world. Jobs in this field range from organizing the swift and safe deployment of troops to destinations halfway across the world; to purchasing and keeping inventory of food and living supplies, equipment, and ammunition; to delivering mail to troops and their families. There were 218,423 soldiers employed in transportation, supply, and logistics occupations in 2005.

HISTORY

As the modes of transportation have improved over the centuries, so have the means of transporting freight from place to place. Businesses can now choose from among many alternatives—air, water, truck, or rail—to determine the best method for sending their goods. They want to find the method that will be the most efficient, economical, and reliable arrangement for each particular type of cargo.

With the rise of mass production techniques in the 20th century, manufacturers have been able to produce more products than ever before. A company may produce hundreds of thousands of products each year, and each must reach its ultimate destination, the

QUICK FACTS

School Subjects
Business
Computer science
Mathematics

Personal Skills
Following instructions
Leadership/management
Technical/scientific

Work Environment
Indoors and outdoors
Primarily multiple locations

Minimum Education Level
Varies by career specialty

Salary Range
$15,617 to $30,618 to $66,154 (enlisted personnel)
$25,358 to $70,877 to $174,103 (officers)

Outlook
About as fast as the average

DOT
184, 185, 290, 292, 905

GOE
07.01.01, 07.05.01, 09.05.01, 09.08.01, 13.02.04

O*NET-SOC
11-3071.01, 41-1011.00, 41-2031.00, 43-5071.00, 53-3031.00, 53-3032.00, 53-3033.00, 53-3099.00

consumer. The vast numbers of products have created a need for people who specialize in seeing that products are packed, shipped, and received properly and efficiently. Today's transportation, supply, and logistics professionals make use of the latest technological innovations to coordinate the shipping and receiving of products, equipment, and other goods worldwide.

Transportation, supply, and logistics workers have been key members of the military for thousands of years—from the Peloponnesian War to the Iraq War. They have played an important role throughout the history of the U.S. military—whether sending muskets and dried foodstuffs to troops during the Revolutionary War; transporting fuel, tanks, and replacement personnel to the Pacific theater during World War II; or ensuring that military hospitals and triage stations in Iraq have the necessary gauze, scalpels, plasma, and other supplies needed to treat the wounded.

THE JOB

Many transportation specialties exist in the military. Some of the more popular career options are described below.

Military cargo—food, supplies, weapons, equipment, and mail—is sent via ship, truck, train, or airplane to many destinations throughout the world. *Cargo specialists* ensure that the supplies reach their destinations safely and as quickly as possible. They pack and crate supplies and load them into different vehicles. They may use forklifts or cranes for larger equipment such as weapons and jeeps. Once the cargo reaches its destination, specialists check the cargo for damages against invoice manifests.

Logisticians procure and distribute materials, supplies, and equipment for the military. They negotiate prices with different vendors and suppliers, and place orders to keep inventories at their proper levels. Logisticians also make decisions regarding the transportation and distribution of these supplies. They anticipate what materials and supplies will be needed for each military installation or mission.

Petroleum supply specialists are responsible for the storage, handling, and shipping of fuel and lubricants used by the military. Using hoses, valves, pumps, and forklifts, specialists load oil, fuel, compressed natural gas, various lubricants, and petroleum products onto tankers, airplanes, ships, and trains for transportation. Their duties include testing the fuel for impurities, repairing and maintaining pipeline systems and other equipment, and keeping records for storage and shipping.

It's important for routine maintenance to be performed to prolong the life of costly items such as machinery, aircrafts, and vehicles that range from jeeps to amphibious tankers. *Preventive maintenance analysts* keep military equipment on track by establishing maintenance schedules. They check maintenance schedules and notify mechanics about needed services, monitor maintenance activities, and record them properly.

Military personnel are able to purchase food items and merchandise at military-operated retail stores located on ships or at military bases throughout the world. *Sales and stock specialists* operate these retail stores. Their many duties include ordering and receiving inventory, as well as pricing, inspecting, and displaying the items. They also operate cash registers, tally the daily sales amount, and prepare and record bank deposits.

The military offers different retail stores ranging from commissaries to barbershops for the convenience of service people. *Store managers* are needed to keep these retail stores operating smoothly. They keep track of inventory, plan and implement the store's operating budget, and train and supervise sales personnel.

Supply and warehouse managers anticipate the military's need for supplies and equipment such as food, medicine, and various items needed for daily living. In the Coast Guard, they are referred to as *comptrollers*. Supply and warehouse managers are also responsible for training and supervising personnel who inventory, store, and sell the merchandise and equipment. One of their most important duties is developing and overseeing a procurement budget, some totaling millions of dollars. They evaluate bids from many different vendors and suppliers, and make final decisions on how military money is spent.

The military uses many forms of transportation—airplanes, ships, land vehicles, trains—to move troops, support personnel, supplies, and equipment to military bases, installations, and missions worldwide. This herculean task is overseen by *transportation managers*. They first determine the fastest, most cost-efficient way to move people and materials. Then they schedule the delivery and pickup of supplies and people, sometimes coordinating with commercial shippers or transportation services of host countries. Managers direct team members in packing and crating cargo, especially items that need special attention such as fragile medical supplies or volatile weapons and explosives.

Transportation specialists make travel arrangements for military personnel and cargo. They assess the travel needs of every passenger and determine the type of transport needed—sea, air, or land—documents, and schedules. They also handle the transportation arrangements, packing, and inspection of cargo. Transportation specialists serve as flight and gate attendants for military airplanes, where they check in passengers and their baggage.

Rank and Military Branch By Occupation

Job Title	Rank	Military Branches
Cargo Specialists	Enlisted	Air Force, Army, Coast Guard, Navy
Logisticians	Officer	Air Force, Army, Coast Guard, Marines, Navy
Petroleum Supply Specialists	Enlisted	Air Force, Army, Coast Guard, Marines, Navy
Preventive Maintenance Analysts	Enlisted	Air Force, Army, Coast Guard, Marines, Navy
Sales and Stock Specialists	Enlisted	Air Force, Coast Guard, Marines, Navy
Store Managers	Officer	Air Force, Army, Coast Guard, Marines, Navy
Supply and Warehousing Managers	Officer	Air Force, Army, Coast Guard, Marines, Navy
Transportation Managers	Officer	Air Force, Army, Coast Guard, Marines, Navy
Transportation Specialists	Enlisted	Air Force, Army, Coast Guard, Marines, Navy
Vehicle Drivers	Enlisted	Air Force, Army, Coast Guard, Marines, Navy
Warehousing and Distribution Specialists	Enlisted	Air Force, Army, Coast Guard, Marines, Navy

Source: U.S. Department of Defense

The military needs many different kinds of vehicles to transport troops, supplies, and equipment. *Vehicle drivers* are trained to operate these vehicles, many of which are heavy and challenging to drive such as tank trucks, passenger buses, and semitractor trailers. These drivers often have to navigate their vehicles on rugged roads, singly or as part of a military convoy. Besides safe driving, their duties include care and routine maintenance of their vehicles, making travel routes, and loading cargo.

Warehouse and distribution specialists organize the military's vast inventory of supplies and equipment. This is important since the success of missions may rely on the quick shipment of ammunitions, or the health of soldiers with needed medicines. Their duties include receiving, storing, recording, and issuing supplies. They often use computers to help them monitor incoming and outgoing stock, as well as forklifts and other loading equipment when handling heavier items.

REQUIREMENTS

High School

You can prepare for these careers by taking courses in economics, mathematics, science, and business administration.

Postsecondary Training

Educational requirements in the civilian sector vary by career, but most workers receive on-the-job training or short-term training at community and technical colleges. Managers typically have a bachelor's degree in logistics, business, management, or a related field.

In the military, enlisted personnel (cargo specialists, petroleum supply specialists, preventive maintenance analysts, sales and stock specialists, transportation specialists, vehicle drivers, and warehouse and distribution specialists) train via classroom instruction, on-the-job experience, and advanced course work. Course work varies by specialty. For example, cargo specialists take classes that teach them how to operate and maintain forklifts, cranes, and power winches; load and store cargo; and plan and schedule cargo shipments. They also get hands-on experience by loading cargo. Transportation specialists learn how to plan the transportation of cargo and personnel, handle cargo safely and efficiently, and analyze transportation documents via classroom instruction and on-the-job training. Vehicle drivers take classes that teach them how to avoid accidents, drive heavy military vehicles, understand road signs in foreign countries, and keep their vehicles maintained.

Officers (logisticians, store managers, supply and warehouse managers, and transportation managers) must enter the military with at least a bachelor's degree. Once they join the military, they receive additional specialized training via classroom and practical exercises.

Visit the U.S. Department of Defense's Web site, http://www.todaysmilitary.com, for more on military training in transportation, supply, and logistics careers.

Other Requirements

Enlisted personnel working within this military occupation need to be team players and able to follow orders closely. Certain jobs, such as those of sales and stock specialists, call for much interaction with people. Other jobs such as preventive maintenance analyst require mechanical aptitude. Many of the enlisted positions call for heavy, physical labor. Officer-level jobs—supply and warehouse managers, for example—require individuals with strong managerial skills. They should be able to coordinate large groups of people to complete a project, and work well under pressure or tight deadlines.

EXPLORING

The best opportunity for experience in this field would be a part-time or summer job with a transportation company or a local moving company in a clerical capacity or as a truck helper. In these positions, you would be able to observe the work and responsibilities of traffic agents as well as talk with them about their positions. Other ways to learn about any of the jobs listed in this article include reading books and magazines about these careers and visiting the Web sites of professional associations.

To learn more about career opportunities in the military, visit the Web sites listed at the end of this article.

EMPLOYERS

The U.S. government employs the military. In 2005, 218,423 soldiers were employed in transportation, supply, and logistics occupations: 54,317 individuals served in the air force; 66,565 in the army; 12,799 in the Coast Guard; 29,907 in the marines; and 54,835 in the navy.

STARTING OUT

The best way to get started in this field is by discussing career options with a military recruiter. Visit the Web sites listed at the end of this article to locate a recruiting office near you. To start out in any branch, you will need to pass physical and medical tests, the Armed Services Vocational Aptitude Battery exam, and basic training.

ADVANCEMENT

Each military branch has nine enlisted grades (E-1 through E-9) and 10 officer grades (O-1 through O-10). The higher the number is, the more advanced a person's rank is. The various branches of the military have somewhat different criteria for promoting individuals;

in general, however, promotions depend on factors such as length of time served, demonstrated abilities, recommendations, and scores on written exams. Promotions become more and more competitive as people advance in rank. On average, a diligent enlisted person can expect to earn one of the middle noncommissioned or petty officer rankings (E-4 through E-6); some officers can expect to reach lieutenant colonel or commander (O-5). Outstanding individuals may be able to advance beyond these levels.

EARNINGS

The U.S. Congress sets the pay scales for the military after hearing recommendations from the president. The pay for equivalent grades is the same in all services (that is, anyone with a grade of E-4, for example, will have the same basic pay whether in the army, navy, marines, air force, or Coast Guard). In addition to basic pay, personnel who frequently and regularly participate in combat may earn hazardous duty pay. Other special allowances include special duty pay and foreign duty pay. Earnings start relatively low but increase on a fairly regular basis as individuals advance in rank. See the appendix at the end of this book for detailed information on pay scales for the U.S. military. When reviewing earnings, it is important to keep in mind that members of the military receive free housing, food, and health care—items that civilians typically pay for themselves.

Additional benefits for military personnel include uniform allowances, 30 days' paid vacation time per year, and the opportunity to retire after 20 years of service. Generally, those retiring will receive 40 percent of the average of the highest three years of their base pay. This amount rises incrementally, reaching 75 percent of the average of the highest three years of base pay after 30 years of service. All retirement provisions are subject to change, however, and you should verify them as well as current salary information before you enlist. Those who retire after 20 years of service are usually in their 40s and thus have plenty of time, as well as an accumulation of skills, with which to start a second career.

WORK ENVIRONMENT

The work environment varies according to job specialty. Those dealing with the procurement, storage, or issuance of supplies often work indoors—from offices to storage warehouses to base commissaries. Others, such as travel specialists or travel managers, often find themselves working in airplane terminals and depots, or flying inside an aircraft. Inclement weather is sometimes an issue for those working as vehicle drivers or cargo specialists working on a port's loading dock.

OUTLOOK

Employment in the armed forces is expected to grow about as fast as the average for all occupations through 2014, according to the U.S. Department of Labor. When the economy is good and/or during times of war, more people pursue employment in the civilian workforce, which creates additional opportunities in the military. With the U.S. military involved in several international conflicts, most significantly in Iraq and Afghanistan, demand should continue to be strong for military workers.

Transportation, supply, and logistics workers are vital to the success of the military. Without them, blood, surgical instruments, and other health care supplies would not get delivered to military hospitals; replacement tanks and jeeps would not be sent to front-line units engaged in battle; and food and other supplies would not be delivered to hundreds of military bases throughout the world. As a result, opportunities should be very good.

In the civilian sector, employment in the air transportation industry is projected to grow by 9 percent through 2014, which is slower than the growth predicted for all industries. Employment in the truck transportation and warehousing industry is expected to grow about as fast as the average for all industries.

FOR MORE INFORMATION

For information about the transportation industry, contact

American Society of Transportation and Logistics
1400 Eye Street, NW, Suite 1050
Washington, DC 20005-2209
Tel: 202-580-7270
Email: astl@nitl.org
http://www.astl.org

For general information about route driving, contact

American Trucking Associations
950 North Glebe Road, Suite 210
Arlington, VA 22203-4181
Tel: 703-838-1700
http://www.truckline.com

For information on college programs and to read the online publication, Careers in Logistics, *visit the CSCMP's Web site.*

Council of Supply Chain Management Professionals (CSCMP)
333 East Butterfield Road, Suite 140
Lombard, IL 60148-6016
Tel: 630-574-0985
Email: cscmpadmin@cscmp.org
http://www.cscmp.org

For materials on educational programs in the retail industry, contact

National Retail Federation
325 7th Street, NW, Suite 1100
Washington, DC 20004-2818
Tel: 800-673-4692
http://www.nrf.com

To get information on specific branches of the military, check out this site, which is the home of ArmyTimes.com, NavyTimes.com, AirForceTimes.com, and MarineCorpsTimes.com:

Military City
http://www.militarytimes.com

If you're thinking of joining the armed forces, take a look at this site, which guides students and parents through the decision-making process:

Today's Military
http://www.todaysmilitary.com

For information on military careers, contact

United States Air Force
http://www.airforce.com

United States Army
http://www.goarmy.com

United States Coast Guard
http://www.gocoastguard.com

United States Marine Corps
http://www.marines.com

United States Navy
http://www.navy.com

For information on transportation and supply occupations in the military, contact

U.S. Air Force Air Mobility Command
http://www.amc.af.mil

U.S. Army Surface Deployment and Distribution Command
http://www.sddc.army.mil

U.S. Navy Military Sealift Command
http://www.msc.navy.mil

BASIC PAY—EFFECTIVE JANUARY 1, 2007[1]

Cumulative Years of Service

Pay Grade	2 or less	Over 2	Over 3	Over 4	Over 6	Over 8	Over 10	Over 12	Over 14	Over 16	Over 18	Over 20	Over 22	Over 24	Over 26
O-10[2]												13,659.00	13,725.90	14,011.20	14,508.60
O-9												11,946.60	12,118.50	12,367.20	12,801.30
O-8	8,453.10	8,729.70	8,913.60	8,964.90	9,194.10	9,577.20	9,666.30	10,030.20	10,134.30	10,447.80	10,900.80	11,319.00	11,598.30	11,598.30	11,598.30
O-7	7,023.90	7,350.00	7,501.20	7,621.20	7,838.40	8,052.90	8,301.30	8,548.80	8,797.20	9,577.20	10,236.00	10,236.00	10,236.00	10,236.00	10,287.90
O-6	5,206.20	5,719.20	6,094.50	6,094.50	6,117.60	6,380.10	6,414.60	6,414.60	6,779.10	7,423.80	7,802.10	8,180.10	8,395.20	8,613.00	9,035.70
O-5	4,339.80	4,888.80	5,227.50	5,291.10	5,502.00	5,628.60	5,906.40	6,110.10	6,373.20	6,776.40	6,968.10	7,158.00	7,373.10	7,373.10	7,373.10
O-4	3,744.60	4,334.70	4,623.90	4,688.40	4,956.90	5,244.60	5,602.80	5,882.40	6,076.20	6,187.50	6,252.30	6,252.30	6,252.30	6,252.30	6,252.30
O-3	3,292.20	3,732.30	4,028.40	4,392.00	4,602.00	4,833.00	4,982.70	5,228.40	5,355.90	5,355.90	5,355.90	5,355.90	5,355.90	5,355.90	5,355.90
O-2	2,844.30	3,239.70	3,731.40	3,857.40	3,936.60	3,936.60	3,936.60	3,936.60	3,936.60	3,936.60	3,936.60	3,936.60	3,936.60	3,936.60	3,936.60
O-1	2,469.30	2,569.80	3,106.50	3,106.50	3,106.50	3,106.50	3,106.50	3,106.50	3,106.50	3,106.50	3,106.50	3,106.50	3,106.50	3,106.50	3,106.50
O-3[3]				4,392.00	4,602.00	4,833.00	4,982.70	5,228.40	5,435.40	5,554.20	5,715.90				
O-2[3]				3,857.40	3,936.60	4,062.00	4,273.50	4,437.00	4,558.80	4,558.80	4,558.80				
O-1[3]				3,106.50	3,317.70	3,440.10	3,565.50	3,688.80	3,857.40	3,857.40	3,857.40				
W-5												5,845.80	6,046.50	6,247.50	6,450.00
W-4	3,402.00	3,660.00	3,765.00	3,868.50	4,046.40	4,222.20	4,400.70	4,574.10	4,753.80	5,035.50	5,215.80	5,392.20	5,574.90	5,754.90	5,938.80
W-3	3,106.80	3,236.40	3,369.00	3,412.80	3,552.00	3,711.30	3,921.60	4,129.20	4,350.00	4,515.60	4,680.60	4,751.40	4,824.60	4,984.20	5,143.20
W-2	2,732.70	2,888.70	3,025.50	3,124.50	3,209.70	3,443.70	3,622.50	3,755.10	3,885.00	3,973.80	4,048.80	4,191.00	4,332.30	4,475.40	4,475.40
W-1	2,413.20	2,610.60	2,742.90	2,828.40	3,056.10	3,193.50	3,315.30	3,451.20	3,541.20	3,622.80	3,755.40	3,856.20	3,856.20	3,856.20	3,856.20
E-9[4]							4,110.60	4,203.90	4,321.20	4,459.50	4,598.40	4,821.60	5,010.30	5,209.20	5,512.80
E-8						3,364.80	3,513.90	3,606.00	3,716.40	3,835.80	4,051.80	4,161.30	4,347.30	4,450.50	4,704.90
E-7	2,339.10	2,553.00	2,650.80	2,780.70	2,881.50	3,055.20	3,152.70	3,250.20	3,424.20	3,511.20	3,593.70	3,644.10	3,814.80	3,925.20	4,204.20
E-6	2,023.20	2,226.00	2,324.40	2,419.80	2,519.40	2,744.10	2,831.40	2,928.30	3,013.50	3,043.50	3,064.50	3,064.50	3,064.50	3,064.50	3,064.50
E-5	1,854.00	1,977.90	2,073.30	2,171.40	2,323.80	2,454.90	2,551.50	2,582.10	2,582.10	2,582.10	2,582.10	2,582.10	2,582.10	2,582.10	2,582.10
E-4	1,699.50	1,786.50	1,883.10	1,978.50	2,062.80	2,062.80	2,062.80	2,062.80	2,062.80	2,062.80	2,062.80	2,062.80	2,062.80	2,062.80	2,062.80
E-3	1,534.20	1,630.80	1,729.20	1,729.20	1,729.20	1,729.20	1,729.20	1,729.20	1,729.20	1,729.20	1,729.20	1,729.20	1,729.20	1,729.20	1,729.20
E-2	1,458.90	1,458.90	1,458.90	1,458.90	1,458.90	1,458.90	1,458.90	1,458.90	1,458.90	1,458.90	1,458.90	1,458.90	1,458.90	1,458.90	1,458.90
E-1[5]	1,301.40	1,301.40	1,301.40	1,301.40	1,301.40	1,301.40	1,301.40	1,301.40	1,301.40	1,301.40	1,301.40	1,301.40	1,301.40	1,301.40	1,301.40

Source: U.S. Department of Defense

Notes:

1. Basic pay for an O-7 to O-10 is limited by Level II of the Executive Schedule which is $14,000.10. Basic pay for O-6 and below is limited by Level V of the Executive Schedule which is $11,349.90.
2. While serving as Chairman, Joint Chief of Staff/Vice Chairman, Joint Chief of Staff, Chief of Navy Operations, Commandant of the Marine Corps, Army/Air Force Chief of Staff, Commander of a unified or specified combatant command, basic pay is $15,959.40 *(See note 1 above).*
3. Applicable to O-1 to O-3 with at least 4 years and 1 day of active duty or more than 1460 points as a warrant and/or enlisted member. See Department of Defense Financial Management Regulations for more detailed explanation on who is eligible for this special basic pay rate.
4. For the Master Chief Petty Officer of the Navy, Chief Master Sergeant of the AF, Sergeant Major of the Army or Marine Corps or Senior Enlisted Advisor of the JCS, basic pay is $6,642.60. Combat Zone Tax Exclusion for O-1 and above is based on this basic pay rate plus Hostile Fire Pay/Imminent Danger Pay which is $225.00.
5. Applicable to E-1 with 4 months or more of active duty. Basic pay for an E-1 with less than 4 months of active duty is $1,203.90.

Index

Entries and page numbers in **bold** indicate major treatment of a topic.

A

accounting, budget, and finance occupations 7–13
advancement 10
earnings 10–11
employers 10
employment outlook 11
exploring the field 9
high school requirements 8
history 7–8
information resources 11–13
job description 8
job requirements 8–9
postsecondary training 9
rank and branch by occupation 8
starting out 10
work environment 11
accounting managers 8
accounting specialists 8
Advanced Research Projects Agency (ARPA) 91
aerospace engineers 66
air battle managers 157
aircraft launch and recovery specialists 26
aircraft mechanics 26, 135
Air Force
establishment of 35
lawyer interview 130–133
magazines of 152
airlift pilots and navigators 157
air traffic controllers 26
air traffic control managers 26
armed forces
active duty officers 29
books about 127
countries deployed in 71
earnings 197
enlisted grades 10
enlistment process for 168
magazines of 152
officer grades 160–161
Armed Forces Vocational Aptitude Battery Test 166
armored assault vehicle crew members 36
armored assault vehicle officers 36
Army
magazines of 152
platoon sergeant interview 43–44
Army Air Service 35
Army Corps of Engineers 46
ARPA. *See* Advanced Research Projects Agency
art, design, and music occupations 14–24
advancement 19
earnings 19–20
employers 19
employment outlook 20
exploring the field 18–19
high school requirements 16
history 14–15
information resources 21–22
job description 15–16
job requirements 16–18
postsecondary training 16–17
rank and branch by occupation 17
starting out 19
work environment 20
artillery and missile crew members 36
artillery and missile officers 36–37
audiovisual directors 148
audiovisual technicians 148
Austin, Erica 130–133
automotive mechanics 135
aviation occupations 25–33
advancement 30
earnings 30–31
employers 30
employment outlook 31–32
exploring the field 29–30
high school requirements 27–28
history 25–26
information resources 32–33
job description 26–27
job requirements 27–29
postsecondary training 28
rank and branch by occupation 27
starting out 30
work environment 31

B

boat operators 173
bomber pilots and navigators 157
broadcast directors 148
broadcast journalists 148
broadcast technicians 148
budget managers 8
budget occupations. *See* accounting, budget, and finance occupations
budget specialists 8
building electricians 46
building occupations. *See* construction, building, and extraction occupations

C

cardiopulmonary technicians 78
cargo specialists 188
caseworkers 55

Central Intelligence Agency (CIA) 97–98, 100
chaplains 55
civil engineers 46
civil law 123
Coast Guard
band member interview 23–24
establishment of 34, 173
magazines of 152
Coast Guard ship officers 174
combat mission support officers 37
combat specialty occupations 34–44
advancement 40–41
earnings 41–42
employers 40
employment outlook 42
exploring the field 40
high school requirements 38
history 34–36
information resources 42–43
job descriptions 36–38
job requirements 38–39
postsecondary training 39
rank and branch by occupation 38
starting out 40
work environment 42
comptrollers 188
computer science occupations 90–96
earnings 93–94
employers 93
employment outlook 95
exploring the field 93
high school requirements 92
history 90–91
information resources 95–96
job descriptions 91–92
job requirements 92–93
postsecondary training 92–93
rank and branch by occupation 92
work environment 94
computer systems officers 92
computer systems specialists 91–92
construction, building, and extraction occupations 45–53
advancement 50
earnings 50–51
employers 50
employment outlook 51–52
exploring the field 49–50
high school requirements 48
history 45–46
information resources 52–53
job descriptions 46–47
job requirements 48–49
postsecondary training 48
rank and branch by occupation 47
starting out 50
work environment 51
construction equipment operators 47
construction specialists 47
counseling, social work, and human services occupations 54–62
advancement 58
earnings 58–59
employers 58
employment outlook 60
exploring the field 57–58
high school requirements 56
history 54–55
information resources 60–62
job descriptions 55–56
job requirements 56–57
postsecondary training 56–57
rank and branch by occupation 56
starting out 58
work environment 59–60
counselors 55
court reporters 124–125
criminal law 123
cryptographic technicians 99
culinary services occupations. *See* personal and culinary services occupations

D

deep sea divers 135
dental laboratory technicians 78
dental specialists 77–78
dentists 78
design occupations. *See* art, design, and music occupations
dietitians 78
distribution specialists 190–191
divers 135
drafting technicians 47
drivers 190

E

electrical engineers 66
electrical products repairers 135–136
electricians 46
electroencephalograph (EEG) technicians 78
electronics engineers 66
electronic warfare pilots and navigators 157
engineering, science, and technical occupations 63–75
advancement 70
earnings 71–72
employers 70
employment outlook 73
exploring 69–70
high school requirements 68
history 63–66
information resources 73–75
job description 66–68
job requirements 68–69
postsecondary training 68–69
rank and branch by occupation 65
starting out 70
work environment 72

enlisted grades 10
enlistment process 168
experimental test pilots 156
extraction occupations. *See* construction, building, and extraction occupations

F

field artillery officers 37
fighter pilots and navigators 157
finance managers 8
finance occupations 7–13
finance specialists 8
firemen 173
flight engineers 26
flight instructors 156
food service managers 180

G

graphic designers 15

H

health care occupations 76–89
advancement 84
earnings 85
employers 83
employment outlook 85–86
exploring the field 83
high school requirements 82
history 76–77
information resources 86–89
job descriptions 77–82
job requirements 82–83
postsecondary training 82–83
rank and branch by occupation 81
starting out 84
work environment 85
health services administrators 80
heating and cooling mechanics 136
heavy equipment mechanics 135
helicopter pilots 157
human services occupations. *See* counseling, social work, and human services occupations
Hynes, Michael 43–44

I

illustrators 15
industrial engineers 66
infantry officers 37
information technology occupations 90–96
earnings 93 94
employers 93
employment outlook 95
exploring the field 93
high school requirements 92
history 90–91
information resources 95–96
job descriptions 91–92
job requirements 92–93
postsecondary training 92–93
rank and branch by occupation 92
work environment 94
intelligence officers 98–99
intelligence professionals 97–104
advancement 101–102
earnings 102
employers 101
employment outlook 102–103
exploring the field 100–101
high school requirements 99
history 97–98
information resources 103–104
job description 98–99
job requirements 99–100
postsecondary training 99–100
rank and branch by occupation 99
starting out 101
work environment 102
intelligence specialists 98
interpreters and translators 105–112
advancement 109
earnings 110
employers 109
employment outlook 110–111
exploring the field 108–109
high school requirements 107
history 105–106
information resources 111–112
job description 106–107
job requirements 107–108
postsecondary training 107
rank and branch by occupation 106
starting out 109
work environment 110

J

journalists 148
judges 124

L

law enforcement, security, and protective services occupations 113–121
advancement 118–119
earnings 119
employers 118
employment outlook 120–121
exploring the field 118
high school requirements 115–116
history 113–114
interview with 130–133
job descriptions 114–115
job requirements 115–117
postsecondary training 116
rank and branch by occupation 114–115
starting out 118
work environment 119–120
lawyers 124
League of Nations 105

legal professionals and support occupations 122–133
advancement 127–128
earnings 128
employers 127
employment outlook 129
exploring the field 126–127
high school requirements 125
history 122–123
information resources 129–130
job descriptions 124–125
job requirements 125–126
postsecondary training 125–126
rank and branch by occupation 124
starting out 127
work environment 128
legal specialists 124
life scientists 66, 72
logisticians 188
logistics occupations. *See* transportation, supply, and logistics occupations

M

machinists 136
magazines 152
mapping technicians 47
marine aviation officers 157
Marine Corps
establishment of 34, 173
magazines of 152
marine engineers 67
marine engine mechanics 136
maritime operations occupations. *See* naval and maritime operations occupations
mechanic and repair technologists and technicians 134–145
advancement 141
earnings 141–142
employers 141
employment outlook 142–143
exploring the field 140–141
high school requirements 139
history 134–135
information resources 143–145
job descriptions 135–139
job requirements 139–140
postsecondary training 139
rank and branch by occupation 137
starting out 141
work environment 142
media and public affairs occupations 146–154
advancement 151
earnings 151–153
employers 150–151
employment outlook 153
exploring the field 150
high school requirements 149–150
history 146–148
information resources 153–154
job descriptions 148–149
job requirements 149–150
postsecondary training 149–150
rank and branch by occupation 149
starting out 151
work environment of 153
medical care technicians 78
medical professions. *See* health care occupations
medical record technicians 80–82
medical service technicians 78
merchant marine 172–173
metal workers 139
meteorological specialists 67, 72
Military Entrance Processing Station 168
military glossaries 101
military pilots 155–163
advancement 160–161
earnings 161
employers 160
employment outlook 162
exploring the field 160
high school requirements 157
history 155–156
information resources 162–163
job descriptions 156–157
job requirements 157–160
postsecondary training 157–158
rank and branch by occupation 158
starting out 160
work environment 161–162
military recruiters 164–171
advancement 169
earnings 169–170
employers 169
employment outlook 170
exploring the field 167–169
high school requirements 166
history 164–165
information resources 171
job description 165–166
job requirements 166–167
postsecondary training 166–167
rank and branch by occupation 167
starting out 169
work environment 170
missile crew members 36
missile officers 36–37
morale specialists 180–181
music directors 15
musicians 16
music occupations. *See* art, design, and music occupations

N

NASA. *See* National Aeronautics and Space Administration
National Aeronautics and Space Administration (NASA) 64
naval and maritime operations occupations 172–178
advancement 176

earnings 176–177
employers 175
employment outlook 177
exploring the field 175
high school requirements 174
history 172–173
information resources 178
job descriptions 173–174
job requirements 174–175
postsecondary training 174–175
rank and branch by occupation 176
starting out 175
work environment 177
Naval Construction Force 46
naval flight officers 157
naval submarine officers 174
Navy, magazines of 152
newswriters 148
nondestructive testers 136
nuclear engineers 67, 72

O

occupational therapists 79
occupational therapy specialists 79–80
officer grades 160–161
optical laboratory technicians 78
optometric technicians 79
optometrists 79

P

personal and culinary services occupations 179–186
advancement 183
earnings 183–184
employers 183
employment outlook 185
exploring the field 182–183
high school requirements 181
history 179–180
information resources 185–186
job descriptions 180–181
job requirements 181–182
postsecondary training 181–182
rank and branch by occupation 181
starting out 183
work environment 184–185
petroleum supply specialists 188
photographic specialists 16
physical scientists 67, 72
physical therapists 79
physical therapy specialists 79
physician assistants 80
physicians 80
pilots, military. *See* military pilots
pipe fitters 47
plumbers 47
powerhouse mechanics 136–138
power plant electricians 136
power plant operators 136
precision instrument and equipment repairers 138
preventive maintenance analysts 188
printing specialists 16
protective services occupations. *See* law enforcement, security, and protective services occupations
psychologists 55–56
public affairs occupations. *See* media and public affairs occupations
public information officers 148–149

Q

quartermasters 173

R

reconnaissance pilots and navigators 157
recreation specialists 180–181
recruiters, military. *See* military recruiters
recruiting managers 164
recruiting specialists 164–171
registered nurses 80
religious program specialists 55
repair technologists and technicians. *See* mechanic and repair technologists and technicians
research pilots 156
research test pilots 156
Reserve Officers Training Corps (ROTC) 158
ROTC. *See* Reserve Officers Training Corps

S

sales specialists 188
science occupations. *See* engineering, science, and technical occupations
scuba divers 135
Seabees 46
seamen 173
security occupations. *See* law enforcement, security, and protective services occupations
Servicemembers Opportunity Colleges Program 94
ship electricians 138
ship engineers 174
ship officers 173–174
social workers 56
social work occupations. *See* counseling, social work, and human services occupations
space operations officers 68, 72
space operations specialists 67–68, 72
Special Forces officers 37–38
Special Forces personnel 37
special operations pilots and navigators 157
speech therapists 80
stock specialists 188
store managers 188
submarine officers 173–174
supply managers 188
supply occupations. *See* transportation, supply, and logistics occupations
surgeons 80
surveillance pilots and navigators 157

surveying technicians 47
survival equipment specialists 138

T

tanker pilots and navigators 157
technical analysts 99
technical occupations 63–75
test pilots 156
translators. *See* interpreters and translators
transportation, supply, and logistics occupations 187–196
  advancement 192–193
  earnings 193
  employers 192
  employment outlook 194
  exploring the field 192
  high school requirements 191
  history 187–188
  information resources 194–196
  job descriptions 188–191
  job requirements 191–192
  postsecondary training 191
  rank and branch by occupation 190
  starting out 192
  work environment 193–194
transportation maintenance managers 138
transportation managers 188
transportation specialists 188
turpoprop maritime propeller pilots 157

U

unmanned vehicle operations specialists 68

V

vehicle drivers 190

W

warehouse managers 188
warehouse specialists 190–191
weapons maintenance technicians 138
Weaver, Mark 23–24
welders 139
welfare specialists 180–181